Praise for
Practical No-Till Farming

Andrew Mefferd's new book is a no-nonsense breakdown of
no-till market gardening that gets straight to the point and stays
there. Mefferd's a gifted writer who smoothly communicates not only
the principles that guide no-till gardening but the technical details
in a way that makes you keep reading. *Practical No-Till Farming* is
arguably the most devourable technical manual I've come across.
A great read and an even greater asset to organic flower and
vegetable growers everywhere.

— Jesse Frost, author, *The Living Soil Handbook*

This is a book I wish I had when I was starting out. Even now it helps
clarify what the big deal is about no-till, how it's being defined, and
how folks are actually making it work.

— Josh Volk, author, *Build Your Own Farm Tools* and *Compact Farms*

If you want to start farming but don't think you have enough money
for land and machinery, this could be the book for you. If you're
thinking about transitioning from conventional tillage to no-till
farming, this could be the book for you as well. If you're looking for a
practical, down-to-earth book that explains both the whys and hows
of no-till market gardening, this definitely is the book for you.

— John Ikerd, Professor Emeritus of Agricultural Economics,
University of Missouri-Columbia

PRACTICAL
NO-TILL
FARMING

PRACTICAL NO-TILL FARMING

A QUICK AND DIRTY GUIDE TO
ORGANIC VEGETABLE AND FLOWER GROWING

ANDREW MEFFERD

new society
PUBLISHERS

Cover design by Diane McIntosh. Cover photo: Hillview Farms.

Printed in Canada. First printing November 2022.

Back cover Images: right - Lovin' Mama Farm; middle - Hillview Farm; left - Bare Mountain

Inquiries regarding requests to reprint all or part of *Practical No-Till Farming* should be addressed to New Society Publishers at the address below.

To order directly from the publishers, order online at www.newsociety.com

Any other inquiries can be directed by mail to:
New Society Publishers
P.O. Box 189, Gabriola Island, BC V0R 1X0, Canada (250) 247-9737

Library and Archives Canada Cataloguing in Publication
Title: Practical no-till farming : a quick and dirty guide to organic
    vegetable and flower growing / Andrew Mefferd.
Names: Mefferd, Andrew, author.
Description: Includes bibliographical references and index.
Identifiers: Canadiana (print) 20220273766 | Canadiana
    (ebook) 20220273774 | ISBN 9780865719668 (softcover) |
    ISBN 9781550927603 (PDF) | ISBN 9781771423564 (EPUB)
Subjects: LCSH: No-tillage. | LCSH: Vegetables—Organic farming. |
    LCSH: Flowers—Organic farming. | LCSH: Alternative
    agriculture. | LCSH: Sustainable agriculture. | LCSH: Farms, Small.
Classification: LCC S604 .M45 2022 | DDC 631.5/814—dc23

Funded by the Government of Canada Financé par le gouvernement du Canada | Canadä

New Society Publishers' mission is to publish books that contribute in fundamental ways to building an ecologically sustainable and just society, and to do so with the least possible impact on the environment, in a manner that models this vision.

Contents

PART 2: THE HOW OF NO-TILL

"Land, then, is not merely soil;
it is a fountain of energy flowing through a
circuit of soils, plants, and animals."

— Aldo Leopold, *A Sand County Almanac*, 1949

PART 1

THE WHY OF NO-TILL

Introduction

BECAUSE THIS BOOK is meant to be a quick-start guide, I hesitate to put this introduction between you and the practical stuff. However, it just doesn't seem right to put the *how* of no-till before the *why*.

But, if you are absolutely itching to get your hands dirty right away, you could skip to "Part 2: The How of No-Till." If you choose to do that, I'll bet that once you get going, you will eventually find yourself wondering about the *why* that supports the *how*. If that's the case, you can come back to this section later, when it's of more use to you.

Who This Book Is For

I wrote this book primarily to help people who are interested in no-till but don't know where to start. But it's also for growers who have already started with no-till and are interested in expanding or refining their repertoire of techniques.

Over the past few years, there has been a lot of interest — and a lot written — about no-till. This is a great thing. When I started working on my first no-till farm in 2005, there was very little written information available about the subject. At the time, actually working with farmers who were doing it was almost the only way to learn about no-till.

But, happily, that situation has changed. My first book on no-till, *The Organic No-Till Farming Revolution*, came out just three years ago,

3

in 2019. Since then, at least five more books about no-till have been published. Many more are sure to come.

Interest in no-till is so strong that I am writing this book to fulfill the need that growers have expressed for *a quick-start guide to no-till*. Many volumes' worth of information can and will be written about the relationship between plants and the life of the soil that will add to our understanding of how no-till works, but you don't need an advanced degree to use these techniques. No-till methods are actually really easy to try. To get started, this may be the only book you'll need.

One thing that's different between now and 20 years ago, when I first got interested in no-till, is that people are much more used to the idea of a no-till farm now. Back then, there were plenty of people who were outright skeptical and laughed at the idea of growing crops without tilling. That's the reason I included a lot of grower interviews in *The Organic No-Till Farming Revolution* — so people couldn't doubt what no-till farmers were doing and could hear it from the horse's mouth. Now that people are much more used to the idea of a no-till farm, in this book, I will strip the methods down and talk as simply as possible about how to put them into practice.

As someone who wants to see the local food system grow and flourish, in this book I've focused on methods appropriate for local farmers. However, this is not a book just for already established farmers. I know more than a few farmers who got their start as gardeners, and the ideas in this book can work just as well in a garden as they do on a farm. As well, I very much hope the methods described in this book will inspire many readers to become first-time farmers. So regardless of your scale, good luck reducing or eliminating tillage with these no-till techniques!

Tilling Was Once the Only Answer

Tillage has been a standard in modern agriculture for so long, it's become a paradigm. But what we're learning from scientific research and experimentations on farms is that we don't need to stir the soil on a grand scale with our plows. If we start to think of the life in the soil as *our micro livestock* — actual living beings down there turning the soil on a micro level for us — we can make a revolutionary change in the paradigm. We can replace the need to plow the soil with natural processes. We'll be helping ourselves if we let the little guys do their work for us. I'm pretty sure they'd rather do it than us, and the past 20 years' worth of experience with no-till systems shows they are up to the task.

Enthusiasm and Skepticism for No-Till

When I was apprenticing on farms around the United States — in Pennsylvania, California, Washington State, Virginia, New York State, and finally Maine — all of the farms I worked for tilled. Mostly by tractor but some by horse, one thing they all had in common was plowing. The only exception to that was the Virginia Tech research farm that I worked on in 2005. Even after working there, when my wife and I went on to start a farm, we tilled — just like most everyone we had worked for.

My whole farming career, I've been told that tillage was bad. However, much of the time there was no viable alternative. So on our own farm, we kept tilling. But we were searching for ways to farm without it. With the methods we had available to us at that time, we couldn't figure out how to make no-till work on a small, diverse vegetable farm. What we could've used was a guide to all the various ways of going no-till. If we had had one, we could've selected methods suited to our farm. The guide that I wish we had when we started out is what I've written here.

We know tillage has many disadvantages that we can avoid by choosing a farming system that doesn't involve tillage; the rest of this book is dedicated to helping you find the right system for your farm. The advantages and disadvantages are covered in detail below, but among the big-picture advantages of no-till is the potential to improve growers' lives and the ability to alleviate some of the ills of our time. So it's no wonder there's a growing component of the farming community that is rushing to embrace no-till.

Many of the people I meet already have an opinion about no-till by the time we start talking about it. And, as with most things that have the potential to change paradigms and disrupt systems, a lot of people tend to be on one extreme of the spectrum or the other. In other words, they tend to be either very enthusiastic about no-till or very skeptical. In the discussion below about the benefits and drawbacks of both no-till and tillage, we'll look at the reasons why people are excited and skeptical (sometimes at the same time) about no-till.

Becoming a No-Till Farmer

No-till is like a lot of things in life, in that if you know *what* to do, you can go from beginner to competent practitioner quickly. It's when you don't even know where to start that you can spend a lot of time spinning your wheels. As anyone who has ever bought an electronic gadget knows, the included booklet that has pages and pages of explanation is preceded by a quick-start guide. They know all you really want to do is turn it on.

The process of understanding no-till methods, choosing one or more that are appropriate to your farm, and then getting started with a no-till method is a little more complicated than turning on a new smartphone. But this book will allow you to get growing with just what you need and nothing more. Once you've gotten going, you can get into identifying the species of invertebrates in your soil, come up with cash/cover crop rotations that make the most of your season, and fine-tune the rest of your system. But in the meantime, too much information can be a distraction. Learn to ride the bike before you try to pop a wheelie.

Whether you just heard of no-till and are wondering how anybody could possibly grow anything without tillage, or whether you're staring at a grassy field wondering how you're going to grow something in it, let this be your guide to picking methods and getting started with no-till. If you want to get deeper into any of the methods, check the bibliography for further reading.

Defining No-Till: What Counts as Tillage Anyway?

We should stop here and establish a definition of tillage. At the bare minimum, and for the purposes of this book, tillage means *an action that inverts or mixes soil layers.* Textbook examples of tillage are

Though from above a power harrow may look a lot like a rototiller, the mode of action is different. Instead of tines churning through and mixing soil layers from top to bottom, a rotary harrow has tines that rotate on a vertical axis, loosening the top of the soil instead of mixing layers. When set deep, rotary harrows will still disturb a lot of soil, but when run shallowly, they can rough up the surface just enough to get some loose soil to plant into without deep soil disturbance. CREDIT: PHOTO BY BCS AMERICA

In this view under the hood of a power harrow, you can see how each of the five sets of tines rotates independently, with a vertical "egg beater" motion, instead of the horizontal mixing of a rototiller.
CREDIT: PHOTO BY BCS AMERICA

moldboard plowing, where the soil at the bottom of the plow is flipped on top of the top layer, and rototilling, where the soil is violently mixed by the beaters on a *rototiller.* Different growers may count other things as tillage, such as the use of rotary harrows and other devices that disturb the soil surface without inverting it.

New Adaptation for Old Methods

There's nothing new under the sun — or under the soil, for that matter. I did not make up any of the methods described in this book; versions of no-till have existed under various names over the years: *lasagna gardening; no-dig; Ruth Stout's year-round mulch method;* and many others. Going back further, planting methods without tillage were used by the Incas, ancient Egyptians, and many Indigenous cultures over our 10,000-year agricultural history.

What *has* changed, is the interest in making no-till methods efficient on a farm scale. This is partially because of our increased understanding of the complex and crucial role of soil life both in feeding plants and in keeping the soil healthy for the long term. I examine some of this in the section below, "The Power of the Soil."

The Promise of No-Till

One "proof of concept" that no-till has advantages for the grower can be seen in the rapid adoption of no-till in conventional row crop farming. Experiments in the US with no-till field crops on a large scale began in the 1970s, but they really took off when genetically modified corn and soy varieties were developed in the 1990s. After that, a large percentage of the enormous acreage of those crops grown in North America quickly went no-till. "Data from the Agricultural Resources Management Survey on the production practices of corn, cotton, soybean, and wheat producers show that roughly half (51 percent) used either no-till or strip-till at least once over a 4-year period."[1] Considering that those are some of the most widely grown crops in North America, half of just corn and soy would add up to over 100 million acres in the US.

But conventional row crop no-till farming also involves technologies that many farmers are loathe to adopt: it gets around the problem of weeds with genetically modified crops that can withstand herbicide application. This has led to increasing amounts of herbicide usage even though the amount of tillage has gone down. As poor of a trade as exchanging tillage for toxic chemicals is, it is proof of concept that, even on the grandest scale, cutting out tillage can save growers time and money. But it is reliant on methodology no organic grower would want to emulate. So the challenge becomes, how to reap the benefits of not tilling without chemicals?

It's difficult to say whether the widespread adoption of conventional no-till practices is an overall win or a loss for the environment. Although many tout the environmental benefits of reduced erosion in conventional no-till farming, "globally, glyphosate [the herbicide known best as Roundup] use has risen almost 15-fold since so-called 'Roundup Ready,' genetically engineered glyphosate-tolerant crops were introduced in 1996."[2] Also, it has been found that "the concentration and the load of pesticides were greater in runoff from no-till fields than conventional fields."[3]

So, considering that conventional no-till has led to increased herbicide usage, and it also leads to increased pesticide runoff, conventional no-till is *not* a model for a healthier environment.

Obviously, organic no-till requires an approach that does not include herbicides or genetically modified crops. As will be discussed in more detail below, no-till provides many opportunities for a more organic approach. Not only does no-till have the potential to save time and money, but it also builds organic matter, sequesters carbon, increases water infiltration and water-holding capacity, and improves soil life. So it would appear to be a remedy for a lot of the ills of farming of our time, including erosion, drought, and climate change.

Whether or not you're certified organic, the methods in this book, and organic no-till in general, work because of *healthy soil*. And as much as reducing tillage is a step toward healthy soil, spraying chemicals like the herbicides that make conventional no-till possible is a step backward because it kills soil life that may have been spared by the plow. Understanding what makes healthy soil and how it can be supported is a vast subject; what we have yet to learn about the soil alone could fill many books. So we will summarize what we know as it relates to no-till in the next section.

THE POWER OF THE SOIL

I LIKE THE QUOTE FROM ALDO LEOPOLD that appears at the beginning of this section because it describes how soil functions as both *battery* and *transmission* for the energy from the sun that is converted into storable energy by plants. For so long, Western agriculture has viewed soil as a passive thing that is just there to anchor plants, when in truth there is a lot more going on down there. The life in the soil biome has evolved many symbiotic relationships with plants; it processes nutrients in the soil in exchange for some energy from the sun.

Looking at it the other way around: plants take energy from the sun and share some of it with the life in the soil. In exchange, soil life has many ways of sharing resources in the soil with plants. This is symbiosis. Plants are the link between the extraterrestrial energy from the sun and the terrestrial resources in the soil, and it is these relationships that make life possible on Earth. That may sound like a grandiose claim, but the vast majority of all life on Earth gets its energy from the sun, directly in the case of plants, or indirectly in the case of herbivores and carnivores.

The life in the soil is both creator and destroyer. It helps plants grow and then breaks down what's left when plants die. This ability for the soil life to break down dead matter is no small thing. It has transformed life on Earth. For example, coal formation declined around 300 million years ago because that's when fungi evolved the ability to break down lignins (the component that protects tree cells) that trees had developed 100 million years prior to protect themselves from being eaten.

Before fungi evolved the ability to break them down, trees didn't decay; they would simply drop where they died, and — eventually — some became coal. But once fungi developed the ability to break down lignins, trees started rotting, and their stored energy became available to other life forms. As growers, we can put the power of decomposers to use — if we just stop killing them with tillage, pesticides, and overfertilization.

So, the soil is the great recycler for our planet. It breaks down the organic matter from creatures fed by the sun; this is why we say when something has died that it is going back to the Earth. Once microbes and other creatures have broken dead organic matter down, the soil

Think of the Earth as an organism with its stomach on the outside. Since the Earth is filled with molten rock and is closed off from the outside world by its crust, the digesting and releasing of biological energy isn't happening on the *inside* (as with our own bodies and most creatures we are familiar with), it is happening on the *outside*. The entire surface of the planet is a stomach, breaking down, digesting, and regrowing the matter and energy on the surface of the planet.

My wife Ann, who is a geologist, told me she likes studying rocks because they're the Earth's bones; by that analogy, soil is the planet's exterior stomach, held in by the Earth's skin, its vegetation. What you're doing when you put a tarp on the Earth is inviting the soil life, which drives the digestion, up to the surface of the soil to eat what is there. In the dark, the plants die of starvation, and in turn, the microfauna digest the dead plants. That's why as a soil farmer, the soil life is your *micro livestock*.

We used to think in agriculture that chemistry drove biology; what we're realizing is that it's the biology that drives the chemistry, at least when it comes to the availability of nutrients for plants.

becomes a *battery*, storing energy and nutrition from the previous forms of all that organic matter.

We are learning more all the time about how, in addition to being fed by breaking down the organic matter from dead organisms (and keeping the Earth from filling up with dead things), the soil biome is also fed by exudates from plant roots. Thus, energy is passed from the sun through plants and then to the organisms of the soil which, in turn, give some of that energy back to the plants. It is a symbiotic trade that we have only glimpsed the tantalizing potential of.

These all-important relationships were misunderstood by chemical agriculture, which viewed the soil as simply a substrate to give plant roots something to hold. Until recently, from the conventional agricultural perspective, the real work was thought to be done by chemical companies making fertilizers and the farmers adding them to the soil. It turns out that all along the soil could do the work for us.

Misunderstanding and underestimating the power of the soil is as understandable as it is unfortunate. To mid-20th century chemical company scientists, it was easier and more profitable to study plants' responses to different manufactured fertilizers than it was to study the soil life that was naturally cycling nutrients in the soil.

The "green revolution" technologies that are widely praised for increasing short-term agricultural productivity are also responsible for the long-term degradation of the very farmland they are credited with increasing the productivity of — through erosion and soil biocide caused by tillage and chemicals. Reducing or eliminating tillage is one of the biggest ways growers can stop chasing their tails, trying to augment organic matter by adding it in various forms only to burn it up with tillage and send its carbon back into the atmosphere. This subject is discussed in more detail in the section on "Disadvantages of Tillage."

Putting a Face on the Soil

It is easy to think of soil as a pile of inert brown dust. On the contrary, it is *full of* life. Not only that, but it has "the greatest concentration of biomass anywhere on the planet! Microbes, which make up only one half of one percent of the total soil mass, are the yeasts, algae, protozoa, bacteria, nematodes, and fungi that process organic matter into rich, dark, stable humus in the soil."[4]

Soil is so full of life that it's hard to fully comprehend facts like this: "Much more than a prop to hold up your plants, healthy soil is a jungle of voracious creatures eating and pooping and reproducing their way

It's easy to think of soil as homogenous brown dust; however, when we look closely, there's an amazing amount of diversity within. Dicyrtomina ornata, *a type of springtail, is just one of the many thousands of species living in the soil and crop residue.*
CREDIT: ANDY MURRAY

We're in a golden age of learning about soil. Not only have we taken it for granted in the past and not considered it worth studying, but it's been very hard to study because all the action in the soil takes place in the dark, conducted by organisms who, in most cases, are invisible to the naked eye.

toward glorious soil fertility. A single teaspoon (1 gram) of rich garden soil can hold up to one billion bacteria, several yards of fungal filaments, several thousand protozoa, and scores of nematodes...Most of these creatures are exceedingly small; earthworms and millipedes are giants, in comparison. Each has a role in the secret life of soil."[5]

Even though the soil is teeming with life, much of it is invisible to the naked eye. Also, it's dark down there. So it's been only since the advent of modern imaging that we have been able to actually see what's going on in the soil. Happily, though, we're living in a golden age of soil science. Every day, modern microscopy and photography are getting us better acquainted with the critters who do the work. The combination of improving equipment to study the soil and a better understanding of the importance of soil and soil life has yielded a great deal of informational fruit recently. Now, I know we're trying to keep this book short, but since biology is the engine of no-till practices, we must touch on biology as the power behind no-till before we move on to how to grow no-till.

Soil: Where the Microbe Magic Happens

Almost any discussion of no-till methods will get around to talking about soil life at some point. And that's because no-till tries to let the soil biology cycle nutrients to our plants, instead of needing to spoon-feed them fertilizer. Everybody understands the connection between a cow pooping and that poop being used as plant food. If it's helpful, think of the critters in the soil as little tiny cows.

Life in the soil is doing the same things as life on the surface; soil critters are eating each other, and what they poop out is plant food. This is important because, regardless of what your soil test says is in the soil, it's not all available to your plants. One of the major ways nutrients go from being locked up to plant-available forms is by being cycled through soil microorganisms.

Think about it: natural systems don't depend on someone rototilling every few months to keep things productive. Mother Nature does fine on her own, digesting and recycling the nutrients in any given environment. But since we want to harvest high yields of densely planted, heavy-feeding vegetables and flowers, we must do a little extra for our plants to get them to reach the potential we expect as commercial growers.

It's true that we need a higher level of production than might be supplied by just letting nature take its course on our farms, but we can still tilt things in our plants' favor by eliminating competition and preventing weeds. On the fertility side of things, the less we disturb the soil and the more we encourage the right kinds of organisms in it, the more we can rely on them to feed our plants.

Now of course this is all a gross oversimplification; to really talk about the soil food web would take many, many books. See the Bibliography for resources that will help you better understand how energy is cycling from the sun through our plants and back to us.

Taking Care of Our Livestock

As no-till farmers, it is appropriate to think of soil life as *micro livestock*. And our job is the care and feeding of that micro livestock — because, as already discussed, they make nutrients available to plants.

As you will see, there are many ways of caring. For example, let's say you have a dense cover crop that needs to be incorporated into the soil. You could plow it into the soil, and wait two weeks or more for the soil to start digesting the biomass before planting the next crop. But, with a little more time, you could let the soil life do the incorporation for you. For faster breakdown, you could *flail mow* the cover crop (chop it into bits), and then tarp it and let the worms and other soil microorganisms break down the crop residue — so you don't need to plow.

Soil Life Drives the Success of No-Till Systems

Once we understand soil as more than just something to hold plants and fertilizer, we can talk about why *healthy* soils do what they do. To me, it's like magic — in part because we still know so little about it all. It's thought that we've only identified 10% of all the species that live in the soil.

Modern microscopy makes it easier to understand what is going on in the soil by showing us the creatures that live there.

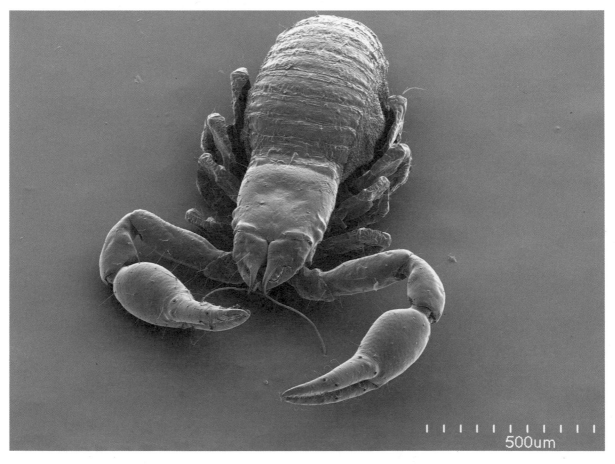

500um

What we do know is that microbes cycle dead plants and other formerly living matter and rocks into usable plant nutrients. Soils that have been tilled and/or heavily sprayed with chemicals have less microbial life. This is one of the reasons many growers trying no-till systems report that they work better over time — most agricultural soils have been degraded through farming, and tilling them less is one of the many steps we can take toward healing them. "Very few people understand that the soil is an ecosystem, so it is our duty to educate as many people as we can that the soil is alive," said Gabe Clark in his book *Dirt to Soil*.

Simply the physical action of tillage kills many of the things that live in the soil. Imagine if you lived in the soil when a rototiller came through; it's harder to imagine surviving going through a rototiller than not. Tillage tends to affect the larger creatures more severely, which makes sense when you think about the long, fragile hyphae of fungi or the chances of an earthworm going unscathed between rototiller tines.

This is also why tilled soil tends to become dominated by bacteria instead of fungi. The smaller organisms like bacteria are more likely to survive, so the surviving bacteria eat the dead and dying after tillage. Imagine a village that is destroyed, and the rats move in to eat whatever

Another interesting critter that we may find in the soil and crop residue is Neanura muscorum. *They are a type of springtail that eats small plants and fungi.*
CREDIT: ANDY MURRAY

is left of the bigger critters. Destroying the village over and over again favors the rats.

This is important because crops prefer a certain ratio of fungi to bacteria in the soil. Most vegetable and flower crops prefer soil that is balanced between fungal and bacterial populations. Many trees, on the other hand, prefer a soil environment that is fungally dominated, like an undisturbed forest floor. When we use no-till methods, we allow important symbiotic relationships to thrive because the soil life is left to function as it has evolved to function. These symbiotic relationships between the plants and the life in the soil can do a lot to help keep our plants healthy — if we let them.

Soil Testing

New approaches to soil testing are showing us that we may not need as much fertility in the soil as traditional soil tests have indicated. Industry standards in testing have all been based on a chemical approach. Testing told us which nutrients were *present* in the soil, but not how available they were to plants. Nutrients can be "locked up" in non-available forms. This means that plants can't get at them to power photosynthesis and growth. When we have active soil life to cycle the nutrients and make them available to our plants, we don't need as much fertilizer as traditional soil tests have been telling us we need.

Currently, a minority of the soil tests done, like the *Solvita Soil Health Suite,* take into account the *biological activity* in soil, which indicates how much of a nutrient might actually be available to plants. Such tests should be helpful for growers to get an idea of how well they're feeding the life in the soil, and they will likely end up giving growers the confidence to use less fertilizer and let the soil feed the plants as the community of soil microorganisms thrives.

The point is, the less we till, the less we will need to till. We avoid compaction from tillage implements and wheels, and we retain the soil's structure, which allows for water infiltration, among many other benefits, which are covered in more detail later. This is a paradigm shift from us being very involved — with tilling, cultivating, and fertilizing — to being able to let the life in the soil keep the soil aerated and the plants fed. And we get a break from a lot of future weeds by simply not churning their seeds up in the first place.

Healthy Soil Can Lead to Healthier Farmers

It's a big mind shift to trust that the soil life can do so much and that we don't need to be intervening with tillage all the time. Even if no-till doesn't end up with you working less (there's always something else to do on a farm), it may mean you can focus more on remunerative activities, cutting out the *muda* (lean-farm terminology for work that doesn't add value), and making your farm more efficient and profitable. Read *The Lean Farm* by Ben Hartman for more on eliminating muda, and many other ideas for simplifying your farm and making it more efficient.

It doesn't matter how straight of a furrow you can plow, or how cleanly you can cultivate; customers don't buy plowing or clean cultivation. You might say those activities are essential to producing something you can sell, but on the other hand, if you could skip plowing and cultivation and have the same amount of salable product, wouldn't you? Even if you never get to the point where your fields are so weed-free that you spend zero time weeding, no-till offers a plan where you can skip those activities and still be profitable; if you choose to work the same amount, a higher percentage of your work can be on directly profitable activities like planting and harvesting.

Left: *A Monobella grassei springtail that lives in the soil and leaf litter. CREDIT: BY AJ CANN, USED UNDER CREATIVE COMMONS ATTRIBUTION-SHARE ALIKE 2.0*

Below: *The soil at Pleasant River Produce on New Zealand's South Island hasn't been turned in over three years. Grower John McCafferty used truckloads of brought-in compost and wood chips to make the transition to no-till. He reports decreased weed pressure and increased water infiltration and retention. Crimped cover crops are visible in the foreground. CREDIT: JOHN MCCAFFERTY*

This picture from my farm shows a bed of sweet potatoes in a hoophouse that was otherwise being tarped with landscape fabric. After removing the previous crop, we simply applied fertilizer and compost, broadforked, and poked the sweet potato slips into holes we had poked for transplanting and firmed the soil back around them. CREDIT: ANDREW MEFFERD

And if you do find that no-till saves you some time, after managing your no-till systems for a few years (because like any other business — but especially with farming — it can take time for systems to become established and smoothly functioning), it's your choice whether to invest any time saved back into the farm, or to use it for family or personal time — which is so important to avoid burnout.

Now that we know these no-till methods work, it is a matter of finding the method that works for your farm. Or, maybe it's a matter of starting a farm from the beginning. In any case, we'll look at the best ways to transition an existing farm to no-till, as well as how to start a farm from scratch without tilling.

FARM SIZE

W E HAVE SO DEEPLY ABSORBED the *get big or get out* mentality of postwar American agriculture that we tend to equate bigger farms with better farms. We need to acknowledge that the only way that farms have been able to get as big as they have is by using chemical and mechanical shortcuts. Some people might look at these "short-cuts" and call them "innovations." But the results are not all good. Many of the mechanical shortcuts resulted in the need for fewer farmers, and many of the chemical shortcuts are negatively impacting human and planetary health, like, for example, the ever-increasing use of pesticides instead of tillage to combat weeds. In this case, chemical methods of control have won out.

Less than 1% of the population are farmers. When less than 1% of the population is feeding the other 99%, it's simply not possible to do conscientious healthy farming. If someone came to you and said, "We can farm super-efficiently, with less than one percent of the population growing the food for the other 99%, but at the expense of environ-mental and human health," would that sound like a good bargain? It doesn't to me. However, it's the situation we find ourselves in.

One of the many reasons that our agricultural policy is so bad (bil-lions of dollars of subsidies is one example) is because so few people have any real connection to agriculture anymore; hardly anyone understands how their food is actually produced. There's food on the shelves when they go to the grocery store, so they assume everything down on the farm is fine. Most people don't understand how bad our current farming system is. Meanwhile, billions of dollars in agricultural subsidies are going to produce food that damages the health of the environment as well as the people who eat it, all for an industry that doesn't really need to be subsidized at all. So, with most people having no real understanding of agriculture, we get bad agricultural policy.

I'd like to advance the radical idea that it would be better if we had *more farmers* growing our food and flowers. The more farmers we have will increase the ratio of farmers to acres, which should not only result in better stewardship of our land and human health but also make the agricultural supply chain more resilient as we diversify the farms that supply it.

Apparently, there are a lot of people out there with this same radical notion. After a century of "get big or get out" agriculture driving market farming to the brink of extinction, the number of farmers markets and other local delivery outlets has rebounded over the past few decades. Meanwhile, many other methods for farmers to sell to the communities they live in have sprung up: CSAs, food hubs, farm stands, and farm-to-school programs, among them. However, despite all this progress, local food and flowers still only comprise a very small percentage of what is consumed — to the detriment of our environment, our communities, and individual human health. More local farmers would be a very good thing, indeed.

Farming as a Career

Most people don't even think of farming as a career. Most Americans don't even *know* a farmer. Part of this is our disconnection from the food chain. Much of this is due to the fact that many American farms are *huge*, and it's hard to imagine where one would get a start. At the same time, I think many people have looked down on farming for a long time.

One of the reasons I think more people should be farmers is because I think many people would *like* being farmers. I'm guessing that a lot of people who never even considered it would enjoy farming if it were presented as a viable career choice. I'd bet you know someone stuck in a cubicle or a retail job right now who would love getting their hands dirty but don't know where to start. One of the values of no-till is that it can allow a person to start their own farm and be their own boss. Though it can come in the form of quitting your job one day and starting a farm the next, I also foresee people going more slowly through the no-till gateway into farming — starting a garden on whatever land is available with a tarp and a few tools, building the soil and confidence in their farming methods, and then taking the leap by applying to get into a farmers market, or starting a CSA.

NO-TILL: A GATEWAY METHOD

T HE ADVANTAGES OF NO-TILL, such as they are, and even taken with its disadvantages, make no-till farming of outsized importance to one important group of farmers: *potential* farmers. No-till growing makes it possible to start a profitable farm with little to no equipment, on a very small land base. Heck, with a big enough yard (or multiple rented yards), you could tarp down your lawn and start a farm right there. It is possible for one person to start a no-till farm on an acre of borrowed or leased land with hand tools and tarps and very little investment.

Starting a small farm is the entry point into farming for so many people who might want to farm but otherwise can't imagine financing a big piece of land, tractor, etc. We should not underestimate the value of small farms. Most big farms *started out* small; everyone's got to start somewhere. And the more people we can get to start small farms, the more small farms that will grow into medium and large farms. Overall, it is my goal, personally and with *Growing for Market* magazine, to increase the number of small farmers and small farms. The supply chain disruptions of the last couple of years have shown everybody what local growers have known for years: a more diversified food system that isn't heavily reliant on a couple of production areas for major crops is a more stable food system.

Currently, some low single-digit percentage of food in North America is grown and eaten locally. Much of the food has traveled 2,500 or more miles before it reaches the consumer, and many flowers travel even further. Local farming isn't a silver bullet against supply chain disruption, climate change, and loss of community, but it has a part in solving those problems. Especially in that it lowers the barriers to entry for new farmers, I think no-till has a part to play in getting the percentage of food that is grown locally out of the single digits. The thing that I would really like to see and encourage is more direct-market vegetable and flower farms selling in the regions they grow in.

Many people have not been able to get into farming due to a lack of access to land and the capital for equipment. The fact that you can start a profitable farm that could support you and your family on a small footprint and with a low investment in equipment means that no-till has outsized importance for new farmers who may be able to start a farm on land that they don't even own and gradually bootstrap themselves into a larger farm. No-till is a gateway technology that can allow more people to try — and, hopefully, stay in — farming, with less of an investment than ever before. It's entirely possible to stock a farmers market stand or deliver 100 CSA shares off of a couple of acres. Whether you deliver the produce by CSA, farmers market, or florist sales, the bottom line is that one person or a few people can make a living off of a couple of acres.

I don't foresee a world developing overnight in which everything is locally grown, but I do foresee one where more than a single-digit percentage of our food and flowers comes from local sources. I hope the supply chain disruptions that we have experienced with COVID, including empty shelves at the grocery store, show people the importance of having a diversity of production and supply for something as vital as food.

Evolution of Your Farm

There's always going to be some soil disturbance in farming; I think it's fair to say that techniques that only disturb the very surface of the soil and don't invert soil layers can be called no-till, whereas ones that disturb the soil deeply with the mixing of layers, like moldboard plowing or rototilling, are obviously not no-till. Shallow soil disturbance with a hoe or rake to create a seedbed for direct seeding is an example of a much less invasive use of soil disturbance than, say, frequent rototilling.

The discussion of what is or isn't no-till aside, I think everything we're learning about soil health being conducive to plant health, and the fact that less tillage promotes healthier soils, is pointing us in a direction away from tillage. But I also know it's up to the individual grower to decide what's best for their land and their farm. So even if someone were to go through the no-till gateway and start a farm, only to get established and then decide that what their operation needs is some tillage after all, I would still look at that as a victory for that person to have gotten started.

Small Can Be Beautiful

I value the small-scale farmer as much as the large acreage grower. I think it's a great thing if one person can make a living feeding others off of one acre. Making it possible to make a living off of a parcel as small as one acre is one of the things that this book is aiming to do. No-till is exciting on the individual farm level because it can allow you to start a farm on a very small footprint that will make you a living relatively quickly. And when I step back and look at the big picture, I'm also excited about no-till because I think it makes it easier for people to start farms; and the more farms that are started, the more that will succeed, especially with scale-appropriate techniques like no-till.

I think it's important to look at all the available methods and not confine our discussion to one particular scale or approach because even on the same farm there may be different no-till management solutions for different crops, different times of the year, etc. Also, it is important to know how to scale up. On my farm, for example, even though our background was in vegetable growing, when we wanted to grow ¼ acre of hemp, it was useful to have the skills and tools to do it.

We've been brainwashed to think of big as efficient. Well, there's small and efficient, too; it just looks different than big and efficient. I would argue that no-till itself is an efficiency strategy in that it eliminates all the plowing, harrowing, rototilling, etc., so we can focus on the things that make us money: planting and harvesting.

So, I write this book with the hope that it will give you enough info to be dangerous with a tarp, and inspire people to try tarping down some lawn, growing some vegetables or flowers, and exploring which of these methods are a good fit on farms everywhere. Here's to your success!

NO-TILL VS. TILLAGE

Tillage: The Agricultural Reset Button

ALL THIS TALK ABOUT NOT TILLING begs the question, why do we even till in the first place? Why is tillage the standard for most farmers growing annual crops worldwide? Well, because you can't grow a carrot in a lawn.

I know that's a bit of a flip answer, but it points to the truth of the matter. We till because nature abhors a vacuum, and almost anywhere that is suitable for growing anything, something will already be growing. Tillage has been the reset button for turning a field or lawn into a plantable piece of land; it prepares the soil for planting by wiping away whatever was there in the first place. You can think of it as a vegetation eraser, turning whatever is growing *on* the soil *into* the soil. Once it's done, you can grow whatever you want in that spot, whether it be a vegetable or a flower. Tillage also breaks up hard soil and makes it ready to receive seeds or transplants. However, the disadvantages to tilling are many, and it is my argument that no-till can accomplish the same things — it just requires a paradigm shift. Using no-till techniques, you can turn a lawn or field into a garden or a farm without the use of heavy equipment (some of which wouldn't fit physically or be within the zoning of many urban properties).

Tillage is a means to an end, allowing you to grow whatever it is that you want to grow, where you want to grow it. That it is *effective* at this goal is pretty much all it has going for it. Because, along with opening up the ground for your crops, it also degrades the soil structure, burns up organic matter, kills beneficial soil life, and takes a lot of time, fuel, equipment, and labor. And tilling results in two self-perpetuating cycles: as it burns up organic matter, you must *add in* organic matter to make up for what was lost; and it brings up weed seeds from below the surface, necessitating more tillage or cultivation to get rid of the resulting weeds. These issues are covered more fully, below.

Tillage: An Ancient Practice We Might Want to Avoid

Some of the earliest plows date back to over 6,000 years ago, and no doubt there were digging sticks and other forms of tillage before that, so tillage is almost as old as agriculture itself. So, if something has worked for so long, why mess with it?

Tillage has been used for all these thousands of years because mechanical tillage was one of the only means to the end of making the ground suitable for planting. But the drawbacks to tillage have been known almost as long as people have been tilling.

The farming industry readily acknowledges that tillage is bad. Everyone from the USDA Natural Resources Conservation Service (NRCS) to the farmers that I learned from have told me that tillage is bad. But, until recently, I hadn't heard a lot of ideas for how to grow things without it.

THE DISADVANTAGES OF TILLING

Breakdown in the Nutrient Cycle

THERE'S AN ADAGE IN ORGANIC FARMING: feed the soil and not the crop. I feel like no-till is the most complete manifestation of that philosophy because, once you feed the soil, there has to be something there to *digest it*. This is something that I didn't understand fully until I started learning about the soil microbiome — that it is the life in the soil that we are feeding when it is said we "feed the soil." We can broadly say that tillage is destructive of soil life.

And when we talk about nutrient cycling — making the nutrients in compost and other amendments available to plants — it is the microbiome (you can think of it as *micro livestock*) of the soil that converts those nutrients from the unavailable forms into plant-available forms. So, this whole idea of feeding the soil and not the crop does not work if there is no life in the soil to cycle the nutrients to the plants.

It is possible to see the breakdown in nutrient cycling in, say, a worn-out farm field that has lost most of its life and organic matter. This field can be made productive in conventional agriculture because the chemical fertilizers are in a plant-available form that don't need microbes to cycle them.

The cycle can also break down when microbes are not active. For example, we found that when we propagated peppers and other early

33

It turns out that organic growers have been right for eons when they've said "feed the soil and not the plant." Or you could think of it as "feed the soil to feed the plant." The underlying principle is the same. If you keep your soil healthy, it will form alliances with your plants and make nutrients available to them.

To oversimplify, plants photosynthesize, making food (stored energy) from the energy of the sun, and they share some of that with the creatures in the soil via root exudates. The creatures in the soil pay them back by living, dying, eating each other, and pooping each other out; that poop is plant food as much as cow poop is plant food. We've gone from thinking that chemistry drives plant nutrition to understanding that soil life will make nutrients available to plants if we let it; especially if we encourage the symbiotic relationship, first and foremost by not killing the life in the soil through tillage.

crops in the greenhouse, we would get yellow leaves on the pepper plants. This happened for many years until we finally figured out it was because our potting soil had been stored outside and was still cold; the microbes had not yet started cycling nutrients when we were transplanting the peppers. Now we pre-warm the soil for a couple of weeks before potting up peppers and other cold-sensitive crops. Simply "waking up" the microbes with a little heat fixed the problems we had had with early-season yellow leaves. We didn't have to change the soil, just our practices.

Weeds

I got interested in no-till when I was working on a big 100-acre organic vegetable farm in 2004. As on many farms, weeds were a fact of life; they were constantly growing and going to seed. We knocked them back with physical tillage. Of course, because it was a certified organic farm, we couldn't use herbicides. The farm had three or four cultivating tractors, and it seemed like one of them was always cultivating somewhere on the farm, due to the never-ending battle against the weeds.

Working on that farm was a great experience; for one thing, it showed me how to scale up an organic farm. Weeds that might be dealt with by hand on a 1–2-acre farm were dealt with by the small fleet of Allis Chalmers model G's. I got a lot of experience driving those and other tractors around, but I didn't get into farming to drive tractors. I am a plant nerd and would rather spend my time dealing with plants, not breathing tractor exhaust. (I realize this is personal, and know that there are some farmers out there that live to fire up their tractors.)

But it is a fact of life that tractors are a depreciating expense, and the time spent driving them and cultivating could be spent on something more profitable. I thought there had to be a better way to deal with weeds than just constantly beating them back. Even growers who love to drive their tractors don't want to cultivate any more than necessary.

The other reason it was great for me to work on that farm is that it pushed my thinking in a new direction. All the time spent driving those tractors around allowed my mind to wander and think about alternatives. Other than the USDA and the Rodale Institute, there were not many sources for information on no-till at the time. As the season was drawing to a close in Washington State, I started thinking about what our next move was.

At the time, my wife (and *Growing for Market* magazine co-editor) Ann and I were working on farms in the summertime and doing desk jobs in the wintertime to save money to start a farm. When I was looking online to find a farm to work on the next season, I saw that Professor Ron Morse at Virginia Tech was working with the Rodale Institute and the USDA to investigate no-till vegetable growing systems; specifically, he was running trials looking at how no-till vegetable yields compared to conventionally tilled organic vegetable yields.

Partly because I'm originally from Virginia, I emailed Ron and asked if I could come down and pick his brain about organic no-till. Ron invited us down, and he was gracious enough to tell us about his research. Then the meeting turned into a job interview, and the following season I ended up working for him at Virginia Tech doing no-till research. Ann and I were a year away from starting our own farm, and the plan was to learn the organic no-till method from Ron over the course of 2005 and then start our farm the following season.

The no-till methods they were using were based mostly on the *roller/crimper method* (see sidebar). Though the methods were effective, when we started our farm the next year, we realized that mulch grown in place did not scale down well to the size of the farm that we started. Mulch grown in place was better suited to larger blocks of crops that are all planted at once, instead of the successions of various market crops that we needed to plant every week to keep the farmers market stand stocked. At the Virginia Tech farm at the time, plantings were in *blocks,* grown more like field crops, with large blocks of broccoli, tomatoes, squash, or pumpkins instead of mixed rows of crops. So, even though it's a good method (we've used a variation of it for growing hemp on our own farm for the last few years), it didn't lend itself to small plantings of different market crops every week throughout the season.

I wish I could say that we were clever enough to have adapted the roller/crimper method to our farm at the time. Alas, we did not. And then we forgot about no-till for a decade or so until I took over *Growing for Market* magazine in 2016 and started getting stories from growers about how they were making no-till work on small-scale direct-market farms that had many successions of different crops over the course of a year.

The *roller/crimper method,* also known as *mulch grown in place,* uses a cover crop that is terminated by rolling it down flat and crimping the stems. Thus, the cover crop becomes a mulch that can suppress the weeds that might otherwise grow and compete with the cash crop. Seeds or transplants are planted through holes or slits in the mulch.

These stories reignited my interest in no-till and made me want to learn as much about it as possible so we could start running our farm with no-till methods. Intrigued by the stories I was reading about farmer-developed, down-scalable no-till systems, I looked for more information to apply to our own farm. At the time, there was not enough information for me to transition to no-till. So, that's when I took the year and a half to visit as many no-till farms as possible so I could write up the state-of-the-art in farmer-developed no-till. That was the origin of my first no-till book, *The Organic No-Till Farming Revolution*.

Hard Work and Time-Consuming

Tilling is hard work. Whether it's via an animal or a tractor, it takes a lot of energy to pull a plow through the soil. Tilling also requires a lot of time, especially considering that most tillage is not a single-pass operation. For example, a fairly common order to do things would be as follows: *moldboard plowing,* then *disking,* then *harrowing,* then *rototilling*, and only then *bed forming.* And there might even be other passes needed before a field can be planted. So, if tillage could be eliminated, in many cases five or more passes and five pieces of equipment could be eliminated.

Tilling is one of the biggest jobs on a farm. It is often the case that the horsepower of the tractor you need to buy is determined by how big of a plow you need to pull. Field preparation uses a lot of labor and energy; a large amount of carbon goes into the atmosphere from tractor emissions. This is a significant fact, considering that some farms may need five or more passes to till and render a field plantable.

It is not my position to pass judgment on what does or doesn't count as no-till. My own description of it is generally *any method that doesn't invert the soil profile.* Some growers go so far as to adopt a "no steel in the field" rule, where they try to limit soil disturbance to nothing beyond the opening up of holes to insert seeds or transplants. Some growers even refrain from growing root crops like potatoes and carrots because harvesting them can be a lot like tillage, depending on what method you use.

I don't judge anybody or their methods because I realize that all farms are works in progress; if a book like this helps you to reduce the amount of tillage that you use, I'd consider that a success. So go ahead and set your own goals, with the understanding that, especially with a new farm, it can take years for all the systems to start clicking and for farm goals to be realized.

Physical, Chemical, and Biological Disadvantages

Soil has three properties that we are most interested in agriculturally: the physical, the biological, and the chemical. Tillage is bad for all of them.

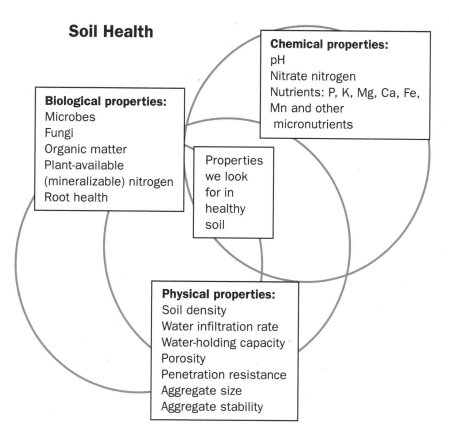

Soil Health

Chemical properties:
pH
Nitrate nitrogen
Nutrients: P, K, Mg, Ca, Fe, Mn and other
 micronutrients

Biological properties:
Microbes
Fungi
Organic matter
Plant-available
(mineralizable) nitrogen
Root health

Properties
we look
for in
healthy
soil

Physical properties:
Soil density
Water infiltration rate
Water-holding capacity
Porosity
Penetration resistance
Aggregate size
Aggregate stability

These are the properties of the soil that we care about as growers. Healthy biology will improve the chemical and physical properties of the soil for our plants. Tillage is bad for all three.
CREDIT: ANDREW MEFFERD

The Physical Consequences of Tilling

The best agricultural soils have *structure*. This is produced when there is a thriving community of microorganisms in the soil. Worms, bacteria, fungi, and other creatures live in the soil; as they move, eat, poop, and die they leave behind voids that become pores for water and gasses to flow in and out of the soil. Tillage eliminates this structure because it crushes, compacts, and pulverizes the soil. Not only does this send soil structure back to square one, but freshly tilled soils are also much more prone to erosion. After tillage, all the soil that you just loosened up, with no plant roots, fungal hyphae, or aggregates holding it together, is susceptible to simply being washed away.

This picture from Hillview Farms in California shows soil with a structure that holds together even after we dug it out of the ground — complete with an earthworm and pores left by various soil critters.
CREDIT: ANDREW MEFFERD

Additionally, all the pores and tunnels that would otherwise carry rain deeper into the soil profile are destroyed, so precipitation tends to pool and run off, taking more soil with it than if it had been absorbed. Considering that we want rainfall to soak in, not run off, tillage is a double whammy, causing less water to soak in and more soil to erode.

Chemical Consequences of Tilling

Tillage burns up organic matter. And by "burning up," I mean the physical process of oxidation. When plants and animals die and decompose into undifferentiated organic matter, the carbon in all that matter is sequestered in the soil. When you expose that organic matter and its carbon to the air through tillage, some of the air's oxygen bonds with the carbon to form carbon dioxide, which then off-gasses into the atmosphere. This is bad for two primary reasons: plants are deprived of the fertility that the organic matter would have provided, and the climate suffers because carbon ends up in the atmosphere instead of staying in your soil.

Organic matter is about 60% carbon, so if you lose even a couple of percentage points of organic matter on an acre, that is a lot of carbon that goes from your soil into the atmosphere. And organic matter absorbs water. A 1% increase in soil organic matter holds an additional 25,000 gallons of water per acre. This is why no-till has an especially big role to play in drought-prone areas; it helps more water seep in instead of allowing it to run off, and it gives the soil a bigger "bank" to hold water in. A very common theme that I heard from growers who had gone no-till was that over time their need to irrigate went *down* as their soils absorbed and banked more water.

Carbon Cycling

Think about how small the seed was that started any given plant, and then imagine all the carbon needed to build that plant to maturity. Plants take carbon from the atmosphere to grow, and they contribute it to the soil when they die and decompose. What we want to do is *sequester* that carbon in the soil rather than cycling it by simply growing plants and then releasing that carbon back into the atmosphere. We want to take all the carbon in the plants we grow and turn as much of it into soil as is possible.

We make deposits into the soil carbon bank by increasing the amount of organic matter in the soil, either through *cover cropping* or through the addition of organic matter in the form of compost or other mulches. If we can gradually bring up the percentage of organic matter in our soil, not only does everything grow better, but we know we're taking carbon out of the atmosphere and depositing it into the soil carbon bank.

Accelerating the oxidation of organic matter by exposing it to the air through tillage promotes a short-term release of fertility — at the expense of the long-term reserves in the soil. As just discussed, the destruction of soil organic matter also releases carbon that had been sequestered in the soil into the atmosphere as carbon dioxide. This is part of the reason why so many growers have a hard time increasing their organic matter levels despite best practices like cover cropping and adding compost.

When we till, we reduce the amount of organic matter in our soil, which means we have to add more in the form of cover crops, compost, or other organic matter just to stay level, keeping the same amount of organic matter we had in the first place. This results in a *one-step forward, one-step backward dance* that makes it very frustrating when you are trying to raise the level of organic matter in your soil. Breaking the cycle of tillage can result in increased organic matter levels, which is a boon to growers who have had difficulty increasing the amount of organic matter in their soils.

Climate Change

It is well-known that agriculture is one of the major contributors to climate change. Tillage introduces unnaturally large amounts of oxygen into the soil. As carbon from organic matter is freed through the decomposition process, carbon molecules bond with the abundant oxygen introduced through tillage to become carbon dioxide rising into the atmosphere. And it doesn't get replenished without effort. Depending on the source, it is estimated that 50–70% of the world's carbon from farmland soils has already been off-gassed into the atmosphere due to tillage. Any amount of farmland that we can convert to no-till has the potential to slow the progression of climate change.

Imbalance of Life

Tillage kills the biology that helps maintain healthy soil. Imagine you are one of the organisms that calls the soil home. If someone drags a plow or rototiller through your home, not only would it actually turn your world upside down, it would likely kill you in the process. There's a quote from the NRCS that I used in my other book, but I love it so much that I will quote it again here: "Tilling the soil is the equivalent of an earthquake, hurricane, tornado, and forest fire occurring simultaneously to the world of soil organisms. Simply stated, tillage is bad for the soil." This quote came from a pamphlet entitled "Farming in the 21st Century: A Practical Approach to Soil Health." It goes on to say that "physical soil disturbance such as tillage with a plow, disc, or chisel plow that results in bare or compacted soil is destructive and disruptive to soil microbes and creates a hostile, instead of hospitable, place for them to live and work."

Tillage is particularly bad for soil fungi because they have long hyphae that act as *fungal roots*, which are easily shredded by any kind of tillage implement passing through the soil. Over time, this promotes a bacteria-heavy soil environment (fewer fungi, more bacteria). A balance of roughly 50:50 between bacteria and fungi is better for most flower and vegetable crops.

The soil food web is almost as much of a frontier as the bottom of the ocean or another planet; it's very exciting because there's so much to learn about the relationships between the creatures that cycle nutrients and make for healthy soil. Advances in technology mean that we are learning a lot about the microscopic creatures we depend on to cycle nutrients for life on Earth. However, we still have a lot to learn, and we have probably only scratched the surface of these relationships that produce healthy soil.

This piece of wood has been colonized on the left by fungus. You can see how the hyphae, which function like fungal "roots," are fanning out to the right to colonize the rest of the wood. Fungi function the same way in the soil, using their hyphae to reach out to locate nutrients, and form symbiotic associations with plants. It's very hard to see hyphae in the soil; they are so fragile that it's difficult to excavate them enough to see them without destroying them.
CREDIT: ANDREW MEFFERD

If you rely on tilling, you rely on the weather in the spring. If fields are too wet in the springtime, planting will be delayed because the fields have to be dry enough to be tilled. Conversely, many growers who have gone no-till have found that they can get on the field earlier in the spring (or even grow year-round where climate permits) by removing the impediment of having to wait until the field is dry enough to start the season.

THE ADVANTAGES OF NO-TILL

Advantages for the Environment

AS ALREADY DISCUSSED, one overarching benefit, from which many of the other benefits of no-till flow, is that it can increase the amount of life in the soil. As we have seen, an increase in soil life is a benefit because soil life helps cycle nutrients in the soil as well as helping to build soil organic matter. These two things are important to the growth of crops.

As our understanding of the soil food web has grown, so has interest in how to promote healthy soils. If soil life is the driver of crop growth, promoting soil health can promote crop growth. In fact, Gabe Brown says, "The fusion of life transforms dirt into soil," which is where the title of his book *Dirt to Soil* comes from.[6]

Fungus: The Perennial Roots of the Soil

Many fungi survive from season to season, and their hyphae can act as a living bridge and connection from the life and nutrients in the soil *to* the crop, and from *one crop to the next* since their networks can remain intact over multiple crops — that is, if we don't kill them with tillage. Because hyphae can persist between plantings, a healthy fungal population in soil can act almost as an extended root system for our plants from the day they're transplanted. An established fungal and microbial network ensures that when you put your plants in the ground, there is an existing support system ready to work with them.

There are *many* symbiotic plant-life support systems like fungal hyphae. Some we didn't even know about decades ago. These systems are one of the reasons why starting a no-till farm on land that has been sprayed or tilled excessively may be harder to do; it takes time to rebuild the soil life that will support crops. But the benefits of rebuilding soil life will accumulate over time. Those support systems of sympathetic fungal hyphae and non-pathogenic microorganisms like bacteria and invertebrates will eventually get reestablished. The life-support systems that will help your plants in a well-functioning no-till ecosystem may just not be there from the beginning.

Soil and Water Can Do Their Jobs

No-till systems preserve soil structure. This is doubly important because 1) soil structure is the habitat for soil biology, which helps with nutrient cycling, and 2) soil structure helps with water infiltration. Increasing water infiltration is important because it means the farmer gets to keep the water that falls instead of it running off, and the combination of reduced water runoff and increased water-holding capacity can make the most of precious water (especially important in arid regions). More water seeping into the ground, instead of running off of it, combined with better soil structure both add up to less erosion, which is extremely important for preserving the long-term productivity of our farmland.

It is surprising to learn that though no-till is held up as one of the most regenerative farming practices possible, much of the research into it was initiated from a commercial standpoint of making farming more profitable by conserving moisture, reducing erosion, and farming more efficiently. It's an idea that has benefits for both saving money and saving the Earth.

Increase in Organic Matter

Increased organic matter, along with improved soil structure, can improve infiltration and water-holding capacity. As discussed more fully above in "Disadvantages of Tillage," tillage loosens the soil and releases fertility at the expense of organic matter. No-till, on the other hand, retains and can increase organic matter more easily.

Advantages for Growers

There are a lot of other advantages to no-till that growers will appreciate. Following are some of the most important.

Set Your Produce Apart

No-till gives you a chance to set your produce apart in an increasingly crowded marketplace. It is the job of direct-market growers to decommodify their food and flowers. We can't compete with the efficiencies and shortcuts of industrial agriculture, so we have to educate our customers about what sets our food and flowers apart.

As it becomes clearer every day that planet Earth is in trouble, people are worrying about climate change; it can be demoralizing to feel like there's nothing we can do about it. People are ready to take some action that helps rather than worsens the situation, so they may be *very* ready to hear about the benefits of no-till. Tell your customers how no-till helps pull carbon from the atmosphere into the soil so they know buying from you is supporting carbon sequestration. Tell your customers about some of the other environmental benefits that stem from no-till growing practices, such as reducing pollution from farm machinery, reducing off-gassing carbon dioxide from tillage, and preserving and even growing the life that keeps our soil healthy. If you can highlight all the advantages that no-till has for the Earth, that's a legitimate way to appeal to people's desire to eat in a way that doesn't destroy the Earth.

Every day, consumers are becoming more conscientious about how their choices affect life on Earth. Tell them that you are farming conscientiously, and you may have a customer for life!

FLEXIBLE FIELD PREPARATION TIMING

No-till can decouple your field preparation from the weather because you aren't constrained by waiting for a good time to plow in your area. If you have time in the fall and know you'll be busy in the spring, you can prep your beds in the fall, cover them with a tarp, and just come back and start planting in the spring. If you're busy in the fall but think you'll have time in the spring, just throw a tarp down in the fall to smother any weeds in the meantime and then come back and prep the beds and plant anytime the following spring.

No-till can give you increased flexibility when it comes to the timing of bed prep. If you know you'll have more time in the fall than the spring, you can prep beds the previous fall as in the picture, drip lines and all, and simply peel back the tarp and start planting when temperatures are right. No-till can also eliminate the risk of a planting delay when a wet spring makes fields too wet to plow.
CREDIT: ANDREW MEFFERD

Efficient Use of Space

One thing especially important on small farms is the efficient use of space. A not-insignificant amount of space is taken up on many farms with roadways and turnarounds that have to be big enough to accommodate a tractor or whatever other equipment is used to work the land. For example, you can't farm fence-to-fence if you have a tractor because you need the tractor to be able to turn around at the end of the road without hitting the fence. Smaller tools mean less of the farm space needs to be devoted to roads and pathways.

Efficient Use of Time

No-till methods can speed up succession planting and result in more growing days on a field because you don't have to wait for tillage to replant. Skipping tillage cuts out a big, time-consuming step — one that can take days to get to, especially if the weather doesn't cooperate. In fact, many of the no-till growers I know harvest and replant beds *the same day*; by doing so, they keep living roots in the soil as much as possible, *and* they make the most of their field space by keeping it cropped as much as possible. Since they don't have to stop between crops to till and bedform etc., many growers will take a crop out in the morning and put another one in in the afternoon. We'll cover how to do this later. For now, the point is that no-till can help you go from

Right: *It really simplifies planting when your soil is not compacted and loose enough so that all that is required is just mulching the pathways and putting some amendments and compost on the beds before planting. In this case, we were only ready to plant half of this hoophouse, so we used a tarp as a placeholder on the other half to keep weeds from growing. This is especially important around the edges, where rhizomatous grasses like to push in from outside the hoophouse. When we built this hoophouse, we put landscape fabric down around the ground posts to discourage grass, but the rhizomes don't give up. To do this, we cut holes in the landscape fabric and put them over the ground posts before the top went on the greenhouse, so there was a foot or two of landscape fabric on both sides of each ground post.* CREDIT: ANDREW MEFFERD

harvesting one crop to planting the next in less time and with fewer steps than with tillage.

Reduced Weeding

Most soil is loaded with weed seeds. Every time you turn the soil, it churns up a fresh batch of weed seeds to sprout. So, if you can kill the already-growing weeds in the top layers of the soil (so they don't set new seed), and you stop turning the soil over, you will stop introducing new seeds to germinate. Once you have exhausted the weed seed bank at the surface of the soil and stop churning up fresh weed seeds, the result should be very low weed pressure. You'll only have to deal with weeds that blow in or otherwise arrive on the surface of the soil.

Lower Barriers to Entry

No-till is the perfect gateway for someone to go from being a serious gardener to becoming a small-scale farmer because it can be set up and efficient in a very small space — potentially without many of the traditional trappings of a farm, like a tractor, barn, etc. It almost doesn't matter how bad the soil is; a person could start a profitable small farm on almost any size tract of land they have access to by building soil up on top of whatever the existing soil is.

The two biggest barriers to someone starting a new farm are usually access to *land* and access to *equipment*. Because no-till farming can be very efficient and profitable on a small footprint, it can reduce the land cost for someone getting started. Another factor that can lower the cost is based on the fact that the better farmland is, the more expensive it is. Since no-till can make it almost irrelevant how bad the soil is, a new farmer can make use of less expensive land when forming beds *above* the native soil. So, these methods may open up land that is more affordable to growers because it is not ideal for commercial farming.

I've talked to many growers who chose to grow on clay or other poor soils or on slopes because it was what they could afford. No-till can allow growers to grow on land that was otherwise considered too steep to be farmable because it makes fields less prone to erosion.

Down-Scalability

As already mentioned, no-till stops the equipment from dictating the scale of the farm and lets the farmer decide what scale to grow on. This is particularly important because people are more likely to start out with smaller farms that can be scaled up later. So, since no-till lowers the barriers to starting a farm, it means we'll have more farms that can start out small and be scaled up to whatever size is desired.

Better Moisture Management

Another benefit of improved soil structure is improved moisture management. And I say "management" instead of "moisture conservation" because no-till can help maximize or minimize the amount of moisture in a field, depending on the need. Tillage dries the soil, so in dry areas, simply not tilling can conserve the moisture that naturally exists.

Conversely, by tarping during rainy seasons, fields can be preserved in a dry condition through wet times of the year. Take, for example, places with a winter rainy or snowy season that brings a lot of precipitation; wet springtime fields can delay farmers from getting to the work of prepping and planting. If a field is tarped in the fall when it is still dry, it can be preserved in that condition through winter rains that otherwise might make a field too wet. When impermeable tarps are removed in the spring, they can reveal relatively dry fields, even in wet areas.

I love how no-till opens up new areas for growing. I got the idea of growing between greenhouses from Ricky Baruc of Seeds of Solidarity Farm. Ricky is a grower who lives up to the farm's motto to "grow food everywhere." The kiddie pool shown here was perhaps not the greatest thing for soil compaction, but it helped hold the tarps down and kept the kids entertained while we worked in the greenhouses. Although blue tarps usually let too much light through to really be ideal for occultation, we folded this one over so there were two layers covering the soil — and it worked! CREDIT: ANDREW MEFFERD

Above: *After the tarp came off, we spread compost between the greenhouses, mulched it with hay, and planted corn, which was very happy there. In our cold Maine climate, I think it enjoyed the extra heat being vented out of the greenhouses, and perhaps the extra water that drained off the structures as well. Since we have such a short season, we germinated the corn in cells because we need to plant it when nights are still cold, which can lead to spotty germination for sweet corn. Because corn seedlings do not like to be in cells for a long time, we planted them within about a week of germination; we simply poked a hole with a digging bar, dropped the corn seedlings in, and then watered them in.*
CREDIT: ANDREW MEFFERD

Above: *Freshly transplanted corn seedlings, planted right through the mulch.* CREDIT: ANDREW MEFFERD

Right: *Be aware that a tall crop can cut down on the sidewall ventilation when grown close to a structure, as this crop of corn was. But in our relatively cold climate, the corn didn't restrict airflow enough to affect the crops in the adjacent greenhouses.* CREDIT: ANDREW MEFFERD

Opens Up New Land for Farming

With tillage, land that could be profitably grown on was limited to land that could be tilled. The significance of this becomes apparent when we consider how much land *can't* be tilled. One of the considerations for rating farmland is whether it can be plowed or not; a lot of land that otherwise has good soil is not considered prime farmland due to being too steep or too rocky to be plowed.

No-till opens up a lot of farmland for the potential of farming, just by taking plowing out of the equation. It also adds some nontraditional farmland back into the mix; for example, it puts suburban yards, urban rooftops, and contaminated sites (for flower production) into play.

Clearly, growing on sites that may have soil contamination requires extra considerations to ensure plant, human, and environmental health — and more space than we have here. For more on growing on contaminated sites, see the article "Brownfield Flower Farms: Cultivating Blooms on Injured Land" from the March 2022 issue of *Growing for Market* magazine at growingformarket.com.

Suburban yards

Depending on the size of the yard, a single one might not be enough to build a farm business. But you've got to start somewhere, and your neighbors might be willing to rent their yards to you once they see what a nice job you've done with yours. One of the great things about no-till is that it means you don't need to live in a rural area or own a tractor to start a farm.

Many farms have been started in urban and suburban areas using no-till techniques. They don't annoy the neighbors with noise and fumes, and because you don't have to cart your rototiller around, you can use all the space in your vehicle for taking produce to market! For a book all about the possibilities of farming on borrowed suburban and urban land, see *The Urban Farmer* by Curtis Stone.

Rooftops

Though it may sound futuristic, more and more farms are being started on the *roofs* of buildings. What may once have been seen as a novelty can now be seen as a legitimate place to start a viable farm business. And while it is more expensive to start a farm on a roof in a city, it also means you're growing in an area with little competition, surrounded by customers.

The rooftop farms that I am familiar with are almost by default no-till, since you're not getting a tractor on a roof. And as long as you use weed free materials, there's no need to cultivate since there's no weed seed bank on a roof. Beds can be built right on the roof, much like building deep compost mulch beds in a field. Hopefully, the weed problem never gets out of hand on top of a building, since you can't till deeply anyway for fear of hitting the roof.

Contaminated sites

Because it is less likely to turn up dust than regular tillage, no-till is particularly well-suited to growing on sites with soil contamination. Just be aware of what is in the soil before you start growing there. Get a soil test. Some of the tests that detect soil pollutants may be more expensive, but don't gamble with your health — or your customers'. Once you've found out that whatever's in the soil won't harm your health, think about what it would do to someone who ate your crops. Some people grow flowers on sites that have contamination that would make produce harmful to human health.

Keep in mind that there may be regulations for any type of growing on sites that are known to be contaminated or are discovered to be contaminated through soil testing. It's important to know the local rules and regulations that apply to growing on contaminated sites. Depending on the contamination, the regulations might make growing on them more expensive. For example, you might be required to install a geotextile barrier between the soil and your growing beds. On the other hand, there may be incentives, such as grants or tax breaks to help establish a business on a contaminated site, so do your homework and be aware of the risks and rewards before starting a farm on polluted ground.

The Latest Research: Increased Nitrogen under Tarps

One of the most exciting things about no-till right now is that more research is finally being done on the subject. We can contrast this with the period two decades ago when I felt lucky to get a job on the research farm at Virginia Tech because it was one of the few institutions at the time studying no-till, along with the USDA and Rodale Institute.

The increase in research is a really important development because it means that no-till is being taken seriously by research institutions, which should speed up the state-of-the-art as researchers start looking at the techniques that have been invented and proven by farmers.

One of the most promising findings has come from a research group involving Cornell and the University of Maine. They have found that where occultation was practiced, the levels of plant-available nitrate nitrogen were higher than in untarped ground. As tantalizing as this finding is, the researchers are not entirely sure what the mechanism is causing the increased nitrate levels. They say:

> Nitrate is a highly soluble and easily leached compound. Tarps likely reduce nitrate leaching losses by rainfall. Longer tarp durations protect the soil from rainfall for a longer period, allowing more nitrate to remain in the soil from the time of tarp application. It is also possible that tarps prevent waterlogging in poorly drained soils following heavy rain events, thereby decreasing the risk of anaerobic conditions and denitrification. Nitrate uptake by weeds in untarped soils could have also reduced soil nitrate relative to tarped soils...Increased soil temperature and moisture under tarps may also promote microbial activity, thus increasing the rate of nitrification in the soil. Soil microbes respond favorably to increased soil moisture and temperature.[7]

The bottom line for me is that, though we don't know exactly why yet, we now know that tarped soils have higher levels of plant-available nitrogen than untarped soils do. I'm looking forward to future research that illuminates why this is so, in order to be able to make more precise use of this finding.

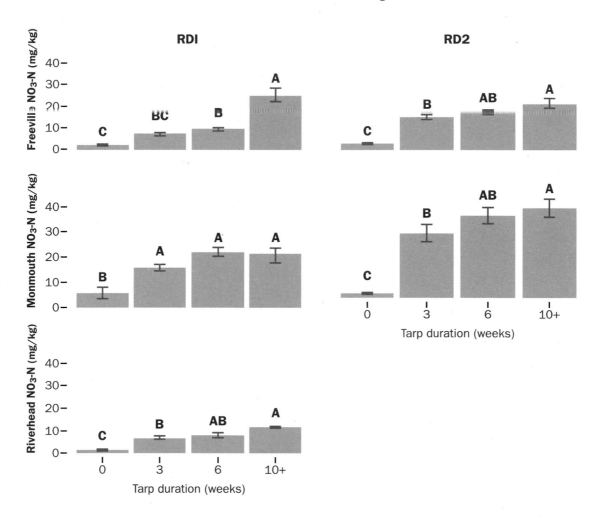

In tarping experiments done by Cornell University and the University of Maine, tarp use of any duration increased soil nitrate (NO3-N) concentrations compared with untarped treatments. Though we need more research to determine exactly why and how tarping works, it is promising that university researchers are starting to study how no-till systems work, so we can make even more effective use of them. For a more detailed analysis, see the article in Hortscience 55(6):819–825, 2020, "Black Plastic Tarps Advance Organic Reduced Tillage I: Impact on Soils, Weed Seed Survival, and Crop Residue." CREDIT: HALEY RYLANDER, CORNELL UNIVERSITY

THE DISADVANTAGES OF NO-TILL

L IKE ANYTHING ELSE, no-till methods have their disadvantages. Here are some of the challenges that no-till systems present.

Fields Are Slower to Warm in Spring

When light-colored mulches are used, fields will be slower to warm in the springtime. So, early-season crops grown on light-colored mulches will probably grow more slowly than those on bare ground or dark mulches. On the other hand, light-colored mulches can be used to advantage in reducing heat stress on crops in hot parts of the year.

Systems Can Take Some Time to Become Established

Chances are, your first year in no-till won't be your best year in no-till. A common theme among the no-till growers that I talked to was that the first year (or two) in a no-till system was difficult, but the systems improved with time as weed pressure went down and the right kind of soil life was encouraged.

Some Methods Are Hard to Scale Up

One of the most frequent questions I hear about no-till is this: how do you scale it up? What works on a small yard or even a field, won't necessarily apply to a bigger scale. Most notably, it is difficult to use tarping or solarization on a very large scale simply because you would need very large tarps — and there's really only so much tarp that can be used before they get too heavy and awkward to handle.

Most small-scale methods work at the scale of a bed or blocks of beds, whereas farmers who are doing acres of production need to work at the field scale, or be able to deal with multiple acres at a time instead of multiple beds at a time. So, the question remains: how to apply the efficiencies that work on a 1- or 2-acre scale to multiple or tens of acres? For growers who are already growing on a larger scale, I can see how the inability to just transition that many acres quickly would be a barrier. At this point, most of the no-till systems that work at field scale suppress weeds but don't eliminate them, so the gradual increase in weed pressure over time in mechanical no-till fields has been a barrier to adoption for larger-scale growers.

Some Pests Can Flourish in High Residue

Many of the systems that use mulch that is left in place during crop growth can result in an initial burst of pests, like slugs and voles. Many of the organic materials that are used to cover the soil, like straw or chopped leaves, can also be good hiding places for pests. However, predator populations such as ground beetles and snakes often catch up and stabilize after pest populations grow. Just be aware that if you already have slugs and voles, there will probably be more of them for a while after adopting systems that leave organic matter residue on the surface of the soil. So have a plan to deal with them until their predator populations catch up.

Perennial Weeds

Perennial weeds can be a problem in any farming system but especially in a no-till system because deeply rooted perennials may be able to survive tarping, or they may have enough energy to come up through even very deep mulches. So, with no way to quickly cultivate weeds, they may become a problem and need to be removed by hand or some other method. Though it should be noted that perennial weeds are problematic for a lot of tillage farms as well, the point is just that with no-till, there's no way to mechanically deal with perennial weeds, especially in the systems that have a lot of residue on the surface to clog up hoes and other mechanical cultivation.

The most realistic thing when it comes to noxious perennial weeds may be to acknowledge that there is no magic bullet solution in no-till, any more than there is with tillage agriculture. In fact, some tillage methods may make some perennial weeds worse; those weeds that spread by rhizomes may actually be chopped up and spread by the act of tillage.

When dealing with pernicious perennial weeds, get rid of what you can by tarping or other no-till methods, but be ready to be vigilant and remove the rest by hand when they rear their ugly rhizomes. If you stay on top of them, you have a good chance of eventually eradicating them. If you acknowledge this and work it into your plan from the start, you will have a better chance of success.

In this photo from Hillview Farms in California, Johnson grass can be seen growing despite being tarped. Occultation may not be enough to kill hardy perennial weeds because they have a lot of energy reserves in their rhizomes. But they are not invincible. You can win the battle if you are vigilant. Credit: Andrew Mefferd

PART 2

THE HOW OF NO-TILL

GETTING STARTED: PRINCIPLES, TECHNIQUES, AND TOOLS

THERE IS NOTHING NEW UNDER THE SUN — or under the soil. The no-till techniques in this book have been written about for years, and have been practiced for much longer than they have been written about. Agriculture predates tillage, so naturally many Indigenous forms of agriculture used methods that did not involve tilling. The methods we will be discussing differ from those that came before in that they focus on the scale and efficiency that is necessary for a farm to be profitable. Even with modern tools and materials, though, no-till relies on many aspects of systems which have been practiced by farmers since the beginning of agriculture. What this book attempts to do is catalog the methods that are a particularly good fit for organic vegetable and flower farms.

Agricultural systems have to do two basic things to produce a crop: provide the raw materials needed for plants to grow in the first place and minimize anything that hinders growth, like weeds, for example. So, first of all, the system has to provide enough sunlight and nutrition for crops so they can grow and yield well. And secondly, a system has to eliminate weeds, pests, and other competition from organisms that want to eat your plants, drink their water, eat their food, or steal their sunshine.

Thousands of agricultural systems have evolved that check these two basic boxes; many of them involve tillage to get rid of weeds,

Get started with field prep as soon as you know you want to grow on a particular piece of land. Regardless of the method you use to get the land ready for planting, the more time you have to prepare, the better the field will be. In this case, we put tarps down on the sod we wanted to turn into an annual crop field. The tarps were put down the fall before we wanted to plant, so they had all fall, winter, and spring to do their work. CREDIT: ANDREW MEFFERD

loosen the soil, and incorporate fertilizer. We are most interested in those systems that fulfill these basics efficiently enough for farms but don't involve tillage.

But before we get into the particulars of the various systems, we should start at the beginning. With your soil.

Start with a Soil Test

Whether you plan on planting directly into native soil or building your soil up, your first step should be to get a soil test. Even if you plan to grow in raised beds, most crop roots will still reach the native soil, so it's worth knowing what's there.

TEST YOUR COMPOST!

If you are planning on using compost, get that tested too. Larger commercial composters should be able to provide you with an analysis of their compost on demand; compost businesses should be testing what they produce on a batch-by-batch basis. If you are buying from a commercial composter and they can't tell you what the analysis is, that's even more of a reason to find out yourself! If you're getting test results from a supplier, make sure it is from the same lot of compost that you are buying.

We frequently buy compost from friends who make compost along with other farm businesses, so as smaller producers, they don't always have a test on hand. Even though we trust them because we know them, we still get it tested. If you didn't make the compost yourself, you really have no idea what went into it; a compost test can tell you whether it's worth paying for.

Poorly made compost can actually harm plants; testing can keep you from paying for it in other ways down the road. Compost can be a significant farm expense, but if it kills your plants, that's even more expensive. Bad compost can kill plants if it isn't entirely finished composting, or is excessively high in salts, or was made with grass clippings or hay that was sprayed with herbicides that survive the composting process.

Most places that do soil testing also do compost testing. Even though a compost test tends to be a little more expensive than a soil test, it will more than pay for itself. Getting a compost test is just like getting a soil test—you put some in a bag and send it off to a soil testing lab. It doesn't cost very much, turns around pretty quickly, and will prevent you from spending your hard-earned money on subpar compost. Most compost tests will tell you how thoroughly composted or "done" the compost is, so you know whether it's really ready to use, or whether it needs to rest a while before you can plant into it. This is especially important if you're applying a lot of compost, as with the establishment of a deep compost mulch system.

You should do soil testing *before* buying or leasing land. Soil tests don't cost much, and without one you really have no idea how good or bad the soil is. Soil tests can explain a lot; for example, if your crops grow fine at first, but then do poorly, knowing that the underlying soil is not good might explain why.

Traditional soil tests tell you how much fertility is in the soil, but they are *not* a good indicator of how much of the fertility is actually available to plants. There are some newer soil tests (like *Solvita* and other *soil respiration tests*) that can measure biological activity and give you a better idea of how much nutrition is actually available to your plants. It's worth checking these out. And you should test every year. Since no-till will make your soil healthier and more able to actively cycle nutrients, you should see a positive change over time.

Clearing the Deck: Getting the Soil Ready

The cycle from season to season and from cash crop to cover crop is an infinite one; we could start anywhere in the cycle. But let's start right in the center, where the cash crop and cover crop cycles intersect: at field prep. If nothing is growing where you want to plant (which could be the case if you're growing on a rooftop or you've already gotten rid of the vegetation), you can go ahead and skip to the section "Getting Started

Right: Flail mowers are a good choice for mowing tall cover crops because they can handle more biomass without getting bogged down like many rotary mowers. This one is mowing down some kale that went to flower. The only time you wouldn't want to use a flail mower is when you want the cover crop to stick around as long as possible as mulch. Flail mowers chop vegetation up into such small pieces that they break down quickly.
Credit: BCS America

Right Insert: In this view of the underside of a flail mower, you can see how it works. As the black cylinder rotates at a high rate of speed, the black V-shaped flails chop anything they encounter into bits.
Credit: BCS America

and Cropping Strategies." If you are planting into an established field with only moderate weed pressure, you can skip down to the section "Crops to Focus On."

If, however, you want to plant where there is established sod or a jungle of weedy plants, you will need to start by wiping the slate clean — getting rid of whatever is growing so you can plant what you actually want to grow. What you want to do is find a method that will allow you to plant your crops without having them be swallowed by grass and weeds. Start with a replacement for tillage. Your main options for wiping the slate clean without tillage are to either cover the field with a tarp to smother what vegetation is there and deprive it of sun (*occultation*) or kill it with too much heat and sun (*solarization*), or to plant a cover crop densely enough to smother weeds and form a mulch when it dies.

Though in theory, you can plant into sod with a *no-till drill*, most people will try to get rid of sod before planting a cover crop into it. Even growers who rely on a dense cover crop as grown-in-place mulch

No-till drills are similar to traditional direct seeders, commonly called seed drills, which is where the name "no-till drill" comes from. The main difference that enables them to function in high-residue environments (with remnants of the previous crop on the soil surface) is the presence of a colter in front of where the seed goes into the ground to cut through the residue, and a press wheel at the back that can firm soil around the seed without getting blocked up with residue.
Credit: USDA NRCS

usually kill the sod first because even a very thick cover crop isn't usually enough to get rid of sod quickly.

With tarping, your mileage will vary quite a bit depending on how long you have to keep the tarps down and what your end goal is. If you just want to make sure a cover crop that you've already rolled down and crimped is dead, you may be able to leave that tarp on for just a couple of weeks during the growing season. However, if you're trying to kill sod or completely digest vegetation on the surface of the soil, that will require more time.

In this picture from the author's time working at the Virginia Tech Kentland research farm, a multi-species cover crop has been no-till drilled into the residue from the previous crop.
CREDIT: BRINKLEY BENSON

In this picture from the research plots at the Virginia Tech Kentland research farm, this multi-species cover crop mix has closed canopy to smother weeds that may be germinating. It will provide a thick layer of biomass to serve as mulch once it is roller/crimped to continue suppressing weeds when a cash crop is planted later.
CREDIT: BRINKLEY BENSON

Opaque tarping (occultation, covered in detail below, under "Tarping") will work much more quickly during the months of the year when plants are actively growing. Since occultation depends on smothering plants and depriving them of light, it doesn't work as fast during the cold parts of the year. However, if you leave tarps down long enough, the soil life will eventually break plants and/or residue down.

If you want the residue from a cover crop to break down more quickly, chopping it up into the smallest pieces possible (with a *flail mower,* for example) will speed up decomposition. As the life in your soil increases over time, cover crop decomposition will speed up too because you will have more microbial mouths to feed that are increasingly active.

If you want the residue from a cover crop to stick around, for example, to act as a mulch to plant through later, ideally you would not mow it at all; the best option is to roll and crimp it in place. Smaller pieces of plant matter break down more quickly, so by leaving the plants as intact as possible, they will last as mulch for as long as possible. This is especially useful for long-season crops, like tomatoes or winter squash, because weeds can start to grow through the mulch late in the season as the mulch starts to break down, and of course you want to prolong the weed-free period for as long as possible.

But if you have to cut your cover crop and still want it to stick around as long as possible to act as mulch, use something that results in minimal chopping, such as a *sickle bar mower* or a *scythe.* Something like a bush hog or lawn mower would be an option

After growing our hemp in the same place for a couple of years, we shifted the tarps down the hill to open up another ¼-acre block to rotate the crop to. The following spring, after almost a year under tarps, the sod was very well broken down. The current crop is growing at the top of the tarp shown in the photo. CREDIT: ANDREW MEFFERD

with a little more chopping, though rotary mowers do have the disadvantage that they tend to blow the mown material around, which is not ideal when you want it to drop in place as mulch. Plus, it can make a mess for crops in the neighboring beds.

For the information you need to make a decision about whether to prepare your field with tarping vs. a roller/crimped cover crop, see the "Tarping" section and the "Mulch Grown in Place" section.

Establishing New Fields

Preparing a New Field (even if it means plowing)

Dare I mention in a no-till book that the first act of some growers who want to establish no-till fields has been to *plow one last time*? As much as this book is written to illuminate the path away from tillage, it's also meant to help people run profitable farm businesses in the real world. And sometimes that means going the long way around instead of going straight at your objective.

So, sometimes you might want to till one last time to get set up for no-till. For example, let's say you need an acre of growing space, but you don't have enough tarps to kill an acre of grass. Or, maybe you have enough tarps, but not enough time for them to sit and completely kill the sod. That means you also don't have enough time to use mulch grown in place to smother the grass.

The speed of plowing means that plenty of growers have used it to wipe the slate clean before starting their no-till journey. As much as I want to encourage a minimum of tillage, if plowing is the only way to get your field ready in time, it's a legitimate choice to do once (or one last time) to get a farm up and running. Especially if there is a high perennial weed population that might be knocked back by plowing once, it's probably worth the one-time soil disturbance to smooth the transition to no-till.

Make the most of it

If you are going to plow one last time, make the most of it. In addition to getting rid of all the weeds at once, plowing is also a way to adjust the nutrients deep in the soil — and to do it more quickly than with no-till methods. So, if there are serious nutrient imbalances deep in the soil profile, make use of that one last plow to incorporate the nutrients your plants need. This is also a great opportunity to incorporate organic matter or amendments deep in the soil, where it will be hard to put them in the future.

There are basically two ways to go when establishing a field. Making the decision on which way to go is a matter of taking into account your farm circumstances and your own personal management goals. Some growers say: "My soil is bad and there are a lot of weeds, I'll plow one last time to kill weeds and incorporate nutrients, and then never again!" And another grower might say: "My soil may be poor, so I'll just make beds on top of the soil." Again, I'm not here to judge. I think growers should do whatever is in the long-term interest of their farms. In most cases, I think it is best to minimize tillage, but if that means tilling to get started, then do what you've got to do. Hopefully, you'll be in business as the steward of a small piece of our Earth for a long time, and all the care you put into it over the years will make up for a little tillage done to get started.

At the end of the day, it comes down to a combination of personal choice and personal circumstances. Get a soil test, figure out how much of which amendments your soil needs, and think about how to incorporate them. Would they be more impactful stirred into the soil by a plow, or layered on top of the soil? If you're going to plow, can you plant a cover crop first to add even more organic matter and nutrients to the soil? I advise you to take the time to gather the information you need to make the right decision for you, your farm, your goals, and your land.

Raised Beds vs. Growing on Flat Ground

No-till can be done equally well on flat ground or in beds that are built up a few inches above the pathway soil level. Many people have been successful with one, or the other, or a combination of the two. Use the situational factors of your farm to choose whether to grow in raised beds or on flat ground. Namely, if you live in a very wet climate or have a high water table, raised beds may help keep the roots of your plants from getting waterlogged; in a very dry climate, planting on flat ground may help access and conserve more moisture. Raised beds may warm up more quickly, whereas flat ground may stay cooler longer.

An advantage of raised beds is the ability to build up your soil. If your native soil is poor, you can put raised beds right on top of the existing soil without having to deal with underlying soil that might be nutrient deficient and/or hard to work. Instead of trying to mix good stuff (compost, soil amendments, etc.) into bad stuff (clay, poor soil,

TRANSPLANTING VS. DIRECT SEEDING

Just as with the question of raised beds vs. growing on flat ground, the question of whether to transplant crops vs. direct seed them is subject to a few considerations for no-till.

While there is no such thing as a specialized "no-till transplant," there are some specific ways that transplants fit into no-till systems. One of the regular advantages of transplanting is that transplants can take root in rougher ground than most seeds want to sprout in. Some no-till farms use transplanting almost exclusively because transplants can thrive in soil that hasn't been finely tilled. Seeds depend on good seed-to-soil contact to germinate, which may not be provided by rougher ground.

So, instead of feeling like you have to create a fine seedbed for seed germination, you can get around the need for fluffy soil by transplanting. If you have a transplant-heavy production system, it can pare bed flips down to a three-step process of 1) removing the old crop, 2) putting on any necessary amendments or organic matter, and 3) transplanting the new crop.

Another thing many no-tillers like about transplants is that they can reduce the amount of time that the soil remains without living roots. Since root exudates are such an important food source for soil life, many no-tillers use a lot of transplants to try to keep living roots in the soil as much as possible.

rocks), with really bad soil it may be easier to just build a bed with good stuff on top of the bare ground. You can plant into that instead of trying to mix a bunch of amendments into deficient soil.

Advantages of raised beds

One advantage of raised beds in a no-till system is that once you've got them built, you can maintain them with minimal labor. With no-till, the biggest maintenance job will be the addition of organic matter and soil amendments to keep the beds at the desired height. This eliminates the step and piece of equipment needed to "bed up" after tillage has flattened everything out.

For example, let's say you finished harvesting a bed in the morning, took the remaining crop residue and weeds off the bed, and were ready to replant by the afternoon. If you direct seed the new crop, there won't be any roots in the soil until the seeds germinate, and even then they will be so small at first that they really won't make much of an impact on feeding soil life.

Though it certainly is a benefit to have living roots in the soil as much as possible, there may be some crops that are just not practical or economical to transplant. It is up to the individual farmer to decide how much to prioritize transplanted crops. There may be a crop that is really important to your market that is not practical to transplant, and in that case, all the other soil-life best practices you're doing will outweigh not having living roots in every single inch of soil all the time.

For a long time, one barrier to going completely over to transplants was the outsized importance of cut salad mix. It is one of the most important crops economically for many farms. However, the development of one-cut lettuce varieties, which are cultivars that can be turned into salad-mix-sized leaves with a single cut, made it possible to grow salad mix without direct seeding. See "One-Cut Salad Mix" in "Crops to Focus On."

Not having to make beds after every crop is one of the best efficiencies of no-till. Saving time and doing away with the purchase of another piece of equipment is great, but that's not even the biggest savings: the biggest savings is all the soil life and fungal mycelium that get preserved in the soil, which will help your plants thrive as soon as they get in the soil.

Sod Bustin', No-Till Style

Regardless of how you intend to manage it later, no-till can simplify opening up new ground. When you know you want to grow somewhere, tarp it or cover crop it *as soon as you can*. For example, if you seal the lease on a piece of land too late in the fall to get a cover crop established, it's not too late to throw a tarp on it! Get out there and put a tarp down ASAP. It can be doing its job, killing sod all winter long, and then your job the following spring will be much easier.

Obviously, putting a tarp down is a lot simpler than planting a cover crop, especially if you're dealing with new ground. That's why I say, if a tarp is anywhere in your plans, go put that tarp down as soon as you have the chance. It can be there doing its job every day it is down, with very little investment from you. The longer the tarp is on, the better the ground will be prepared. You can put the tarp down and go worry about the other million things you have to worry about in a farm business.

Dealing with Weeds

No-till growers have two options for dealing with weeds: blocking them out with tarps and/or mulch or physically removing them. In certain instances, both may be necessary. If, for example, you're using mulch to suppress weeds, and weeds still pop up around or through your mulch, you're going to have to remove them by hand or some other mechanical means.

Be aware that some mechanical methods won't work well to remove weeds in a no-till system. For example, any type of mechanical cultivation tends to get more difficult the more mulch there is. Mulch will clog cultivators — whether stirrup hoes or sweeps — so don't plan on doing a lot of mechanical cultivation if you have a bunch of chunky mulch on the beds.

Reduce the Weed Seed Bank *First*

Stale Seedbedding

Like most problems, weeds are best dealt with before they even emerge. The foundation of no-till weed management is built on the principle that most soils are filled with weed seeds, and the sooner you stop turning the soil over (tilling), the sooner you stop bringing fresh weed seeds up into the germination zone at the soil surface. The techniques are simple, but they require advance planning.

Top: *In their simplest form, tine weeders look like leaf rakes, but the tines are round and much less rigid than most leaf rakes. The idea is that the tines barely penetrate the soil surface, and they vibrate as they are drawn over the soil, vibrating small weeds that have just germinated out of the soil.* Credit: Photo courtesy of Johnny's Selected Seeds

Bottom: *More complicated tine weeders can include adjustments for tine angle and the amount of pressure exerted by the tines. Even though it looks like the tines would take out these lettuce plants, they are able to go around more crops than you would think without damaging them. They are so fast because you don't have to look where you're weeding — just draw the weeder over the bed, crop and all, which is why they are a form of "blind cultivation"; you don't have to look where you're going.* Credit: Photo courtesy of Johnny's Selected Seeds

In the *stale seedbed technique,* weed seeds in the soil are deliberately germinated so you can then kill them before a cash crop goes in. A "stale seedbed" means weed seeds are no longer "fresh" and able to germinate. Most no-till methods are basically stale seedbed techniques applied to an entire farm. Stale seedbedding is one of the best ways of dealing with weeds since, instead of having to work around your crop to get rid of the weeds, you can just kill all the weeds in a whole bed before the crop goes in. But you have to plan ahead; stale seedbedding doesn't work if you don't finish prepping your field with enough time for the weeds to germinate *and then be killed* before the cash crop goes in. Fast transitions of prepping a bed in the morning and planting in the afternoon obviously don't leave enough time to pre-germinate and kill the weeds before the cash crop goes in.

There are multiple ways to make a stale seedbed. *Occultation* (covered in more detail below, under "Tarping") is a very common form of stale seedbedding, especially on no-till farms; the heat and moisture trapped under a black tarp will germinate weed seeds, which then die due to lack of light. Weed seedlings will die more quickly under a tarp during the warmer part of the year, especially if the tarp is dark-colored. This is because the hotter a plant is, the faster it wants to grow; so the combination of being under a warm tarp and trying to grow fast without any sustenance from the sun will cause weed seedlings to exhaust their energy stores faster the warmer it is.

Other ways to achieve a stale seedbed include *flame weeding* and *blind cultivation* with a flex tine weeder. If you know a bed has a lot of weed seeds in it, you deliberately prepare the bed far enough ahead of time so that 1) weeds can sprout, and 2) you can kill them with flame or cultivation before the crop needs to go in. The common theme here is that weeds are deliberately germinated and then killed before the crop is planted.

Check Your Work

On a new farm, you may be unsure of how heavy a load of weed seeds the soil has, and it can be hard to know how successful stale seedbedding has been. So, it's wise to check your work. After a period of occultation, you can pull all or part of a tarp back. If there are a bunch of weeds germinated and still alive, you could just put the tarp back on for a while longer, and the weeds should die from lack of light. How much longer the tarp needs to stay on depends on the weed species and environmental conditions. If you don't see many germinating weeds, it is best to leave the field alone for a few days or a week to see what germinates. You might also want to check your work by taking the time and trouble to irrigate (or just let it rain) where you tarped and see if any weeds sprout. In the best-case scenario, the tarps did their jobs and germinated and then smothered all your weed seeds, and you're left with minimally weedy beds.

But if the best-case scenario doesn't present itself and a lot of weeds still germinate, at least you know they're coming. It may be because your tarp wasn't down long enough to germinate and kill all the weeds, or it may be that the weed species you are dealing with were not conducive to germinating under tarps, or maybe you have perennial weeds that can simply survive the tarping process. There are several ways to try and get rid of them before the cash crop goes in. As just mentioned, you could put the tarp back on for more occultation. This may be a good choice when, for example, spring temps were too cool for all the weed seeds to germinate. Another way of dealing with expected weeds is to put a layer of mulch on beds. This could be a thick layer of compost, wood chips, landscape fabric with holes punched for the plants, or anything else that covers the soil to keep weeds from coming up around the plants. If you are close to planting, you could use *solarization* (using transparent tarps and heat) or various mechanical means to quickly kill off weeds that still germinate after tarping before a cash crop. You can use the knowledge that weeds are germinating to take proactive action and stay ahead of them.

Tarping and solarization are big subjects. They are covered in more detail below, under "Tarping." But, there are several other forms of cultivation that are options when it comes to weeding.

Cultivation

Mechanical cultivation is less common in no-till than it is in conventionally tilled systems for a few reasons. For one thing, some no-till systems leave a lot of residue on the bed. And if there's much of anything on the soil surface, like the residue from a crimped cover crop, for example, it tends to make mechanical cultivation difficult because the mulch clogs up the machinery of mechanical cultivation.

It is one convenience of no-till that you don't need to fluff your soil before you plant into it. In addition to destroying soil life and structure, getting fluffy soil requires extra steps, like plowing, rototilling, bed forming, etc. Not having to go through the whole routine between crops is a radical solution. Farmers can make it work because it gets at the root of two sources of inefficiency on farms: plowing is extra work that also serves to destroy soil life, so eliminating it gets rid of these inefficiencies. If the soil structure is good enough, we can plant into it without needing to fluff it. Of course, the deeper you cultivate into the soil, the more soil life you are disturbing, but many cultivation options can be done without disturbing the soil very deeply.

Flex Tine Weeders

Flex tine weeders look kind of like a leaf rake with wires instead of flat tines; they are a good option when weeds are just germinating, But timing is critical with these since their mode of action is so gentle that they will really only take out weeds when they are at the *white thread stage* of essentially being a single germinated root with not much top growth. You can use flex tine weeders either before a crop goes in for stale seedbedding or when crop plants are big enough for the weeder to go around (but not so big that that they take out the crop plants).

Because the work goes fast with this device, it doesn't require a big time commitment to weed a bed, so you're more likely to get around to weeding promptly at that newly germinated stage. And, especially if you have soil

In this photo from Hillview Farm, you can see how quickly you can move down the bed with a flex tine weeder. In part, this is because you don't have to watch too carefully where you're going. For this to work, the weeds have to be at the white thread stage and crop plants have to be larger but still small enough for the tines to go around the crop.
CREDIT: HILLVIEW FARMS

Left: *Even after crops get too big for a tine weeder to pass over the top, as with this parsley, narrow tine weeders can get rid of newly germinated weeds between rows, like this six-inch model.* CREDIT: PHOTO COURTESY OF JOHNNY'S SELECTED SEEDS

Below: *Tine weeders, like this tractor-scale model made by Treffler, are a great option for getting rid of weeds on a larger scale. The weeder can work on an empty bed as a form of stale seedbedding, to clear a bed with a known high weed-seed bank of newly germinated, thread-stage weeds. It can also work with a crop in the ground as long as the crop is well rooted enough to stay in the ground, and not so leafy that the weeder damages the leaves. The only drawback is that it only works on newly germinated weeds; depending on weed species, a few days after germination, they will be too well rooted to dislodge with a tine weeder.* CREDIT: PHOTO COURTESY OF JEN NOLD

that is prone to crusting over and preventing water infiltration, a nice benefit of the flex tine weeder is that it will break up soil crust so water can penetrate. In the long term, improving soil structure and increasing organic matter should reduce soil crusting, but if you have runoff issues, this type of weeding can help break up that crust so water can soak in.

Flame Weeders

Though flame weeders aren't specifically a no-till technique, they are used extensively on market farms, including by no-tillers. They can make quick work of a weedy bed; it can be convenient to let weeds germinate and then burn them off with blasts of propane. But there is a contradiction between trying to farm as sustainably as possible and being dependent on yet another fossil fuel product — especially one where most of the heat (and emissions) goes straight up into the atmosphere.

Flame weeding might be better than using conventional herbicides, but I would encourage growers to prioritize techniques like solarization or blind cultivation with a flex tine weeder, instead of flame weeding. Blind cultivation doesn't even require any more planning than flame weeding. And a flex tine weeder is cheaper than a flame weeder, and it doesn't involve resupplying with propane. Also, flex tine weeders don't have any chance of exploding if they are used improperly. So, I would encourage anyone considering buying a flame weeder to get a flex tine weeder first; it might just do as good or better a job than a flame weeder, with fewer budgetary and planetary costs. Luckily, there are multiple no-till methods for stale seedbedding that don't involve throwing flame.

Other Cultivation Options

If you don't catch weeds at the thread stage, *stirrup hoes* can be a good option for getting rid of weeds with minimal soil disturbance because they slice weeds off close to the soil surface. However, if you have crop residue or mulch on the soil, it can clog up a stirrup hoe, and the narrower a stirrup hoe is, the more it will clog. (They come in several sizes).

Wire weeders like this one can work well in no-till systems as long as there's not too much residue on the surface to clog them up. They only disturb the soil surface and don't have sharp edges, which can be handy when working around drip tape or other equipment — or in inexperienced hands. CREDIT: PHOTO COURTESY OF JOHNNY'S SELECTED SEEDS

Also, because stirrup hoes can't be mounted on a tractor, this option is limited to cultivation by hand.

There are so many ways to weed, it would be impossible to name them all. Suffice it to say that all manner of sweeps, shovels, basket weeders, or anything else that doesn't cultivate deeply can be used in no-till. However, mechanical cultivation is not often used in smaller-scale no-till, due to the inevitability of clogging up machinery with surface residue and the emphasis on working the weed seed bank down instead of constantly cultivating weeds out. The larger a farm is, the less realistic it is to get rid of every single weed, which is why larger farms tend to rely more on cultivation to get rid of weeds than total elimination of the weed seed bank.

Another reason mechanical cultivation has not been widely used in no-till to this point may be that many of the farms using no-till are on smaller acreages. Though I think as larger growers see the kind of successes that can be had, they will continue researching how to do no-till on a larger scale, which will likely lead to more mechanized cultivation in no-till.

DEALING WITH COMPACTION

I used to think that I had to rototill between every crop or nothing would grow. I realize this was an assumption that was baked into me without any explanation; it was just that every farm I ever worked on did this. Now I can't remember the last time I rototilled in my hoophouses, and everything grows fine. Granted, the soil in our hoophouses is lighter and higher in organic matter than almost any field soil because, as it is such high-value real estate, we've been adding lots of compost and organic matter for years. So now our hoophouses are almost like giant covered pots full of potting soil.

So, you could say we got off easily because our hoophouses got supplied with long-term organic greenhouse soil and that field soils are more in need of loosening — which would be true! None of our fields have soil that is as light with as much organic matter as our greenhouse soil. But still, tilling between every crop just because it's the established routine is a self-perpetuating cycle. Frequent tillage leads to compaction, which needs to be remedied with more tillage. Constantly stirring up the soil to add fertilizer to it disrupts the natural systems that exist to feed plants, necessitating more fertilizer. Stirring up the soil also stirs up weed seeds, which necessitates more tillage to bury the weeds.

When dealing with the issue of compaction, it is useful to think the way a plant thinks. To judge how hard the soil is, we might be tempted to do a "test" by, say, sticking a finger in the soil. But our fingers don't have much in common with the way roots grow, so it's a pretty inaccurate way to estimate how tough of a go of it a root would have in that soil. If you're having trouble trusting that your plants will be able to grow in untilled soil, find out for yourself! Get a penetrometer that will let you know whether your soil is too "tight" for root growth or not. It's a simple device that can be extremely helpful when you are making the decisions to till or not to till.

Sometimes, even when the soil feels hard, it may turn out that it is fine for root growth, and you can skip the step of soil loosening. Or, if your soil is a little tight, you can use less invasive methods like broadforking, a rake, or a tilther to get some loose soil on top of the bed for direct seeding. Or, you could lay some compost on top of the bed and plant into that. We don't have space to go in-depth on the penetrometer here, but I invite you to read the article by Jen Aron in the February 2022 issue of *Growing for Market* magazine: "The Penetrometer: A Simple Tool to Decrease Tillage, Labor and Improve Soil," (growingformarket.com).

In this photo, the author is holding two broadforks. The one on the left, with five tines, is a typical design for soil aeration and loosening. The one on the right, with nine tines, is designed to move soil to lift root crops out of the ground for harvesting. CREDIT: BY MARY HALEY, MXH MARKETING

Left: *If your soil is compacted, you will need to step on the broadfork to get the tines down into the soil. If your soil is not compacted beyond the 300 PSI level that plant roots struggle to grow in, you can skip the trouble of broadforking. If you need to do it, though, start at one end of the bed and broadfork backward, away from where you have already broadforked, so you don't step on the soil you just loosened. Credit: By Mary Haley, MXH Marketing*

Right: *Once you've got the tines of the broadfork in the soil, grab both handles and pull back gently. The action of the tines moving under the soil will help break up compaction. You don't want to pull so hard that the tines lever all the way out of the soil; that would have an effect very similar to tillage. Broadfork every foot or two along the bed, depending on how long the tines of your broadfork are. If your soil is very compacted, you can do more broadforking by leaving less space between broadfork insertions.*
Credit: By Mary Haley, MXH Marketing

TARPING

THERE ARE TWO VERY SPECIFIC ways that tarps are used in no-till. The use of opaque tarps to smother vegetation is known as *occultation*, and the use of clear tarps to burn vegetation out is known as *solarizing*. When no-till farmers (including me) use the term *tarping*, they are most often referring to the opaque tarping that results in occultation.

Occultation

If you've ever been out on the lawn and set a bucket down and got distracted, walked away, and came back a month later, you know that blocking the lawn's access to light and oxygen will kill it and leave you with a perfect little circle of bare earth where the bucket sat. With occultation, we are using this principle — but on a larger scale — to get rid of whatever's growing there in order to make way for a crop.

If there are plants growing where you are tarping, occultation works by 1) starving the plants by depriving them of the energy from the sun, and 2) making it so warm and dark that it speeds up the death of plants under it (this is why dark-colored tarps are recommended). The warmth accelerates plant death because the warmer plants are, the faster they try to grow; so plants under a warm, dark tarp will try to grow fast, and in the process quickly exhaust their energy stores without being able to produce any more via photosynthesis.

Whether there are plants already growing in the area to be tarped or not, the other effect that occultation has is to germinate weed seeds with the warmth under the tarp and then cause them to quickly die when they don't get any light. So tarping can take care of weeds that are already there as well as weeds that have yet to germinate; in the latter regard, it functions as a form of stale seedbedding (covered above), or getting rid of weeds before the crop even goes in.

Occultation takes some time to work, but it can get rid of higher weed pressure and/or more stubborn species of weeds than solarization (see below). It can even break down organic matter if given enough time. In addition to depriving plants of the light that they need to live, occultation also brings the conditions of darkness and protection that usually exist deeper in the soil to the soil surface, which invites soil life *up* to help break down crop residues. That's why, if left in place long enough, tarps can not only kill vegetation that is under them, but they can also allow microorganisms from the soil to completely digest crop residue.

Most growers have found that occultation takes at least four weeks to thoroughly kill the vegetation under it, and longer than that to break it down. Coarse vegetation, for example, a vigorous cover crop like mature rye, will take longer to break down than smaller, less mature plants.

Right: *In this picture from Bare Mountain Farm, you can see how the residue from the previous crop is breaking down underneath the tarp. This picture was taken in the fall at the very end of the growing season. There wasn't enough time left to plant another crop, so putting a tarp down without even removing the previous crop was an efficient way to keep weeds from growing, knock the grass back, and promote the breakdown of the previous crop. If there is still crop residue left when it is time to plant, remaining residue can either be transplanted through or raked off the bed for direct seeding.*
Credit: Andrew Mefferd

Solarization

Solarization refers to putting a clear piece of plastic down on a field to create an intense greenhouse effect and fry weeds. It probably wouldn't be any quicker than occultation on established sod or hay fields to clear new ground, but it can work very quickly for clearing out light weeds after a cash crop.

Regardless of how much we aspire to completely weed-free beds, the reality is that most of us have some weeds in our fields, and it can take years of diligently not letting weeds go to seed before you get a weed-free field. In case this should sound like a made-up, lofty goal, I will say that I have seen weed-free fields with my own eyes. Such

At the beginning of the second year in cultivation, the field looked like this after preplanting solarization. It served to kill most of the overwintered and early-sprouted weeds that would have competed with the crop. As soon as the first field was ready, we moved the tarps to the next one to start preparing it. CREDIT: ANDREW MEFFERD

One of the nice things about solarization is that, since the tarp is clear, you can see your weed-killing progress. However, this can be deceiving; some deep-rooted perennial weeds can survive solarization without showing any signs of life; they may resprout once the tarp comes off.
CREDIT: ANDREW MEFFERD

fields are not unicorns; they really do exist. They had been managed diligently for years to not let a weed go to seed.

However, the reality is that most of us have to deal with more than just the occasional weed, so we need ways to get rid of the weeds following a crop in order to plant the next crop. As already mentioned, one of the primary reasons growers till is to wipe the slate clean and get rid of any remaining weeds before the next crop can go in. This tilling can take many forms, anything from rototilling and re-bedding to completely plowing up a field and starting over again.

However, this is one of those areas where something as simple as a sheet of plastic can replace all that machinery and soil disturbance. With solarization, the idea is that, during the sunny part of the year — when many transitions from one crop to another take place — you can take a piece of clear plastic and put it down on an area that has just been harvested and create an intense greenhouse effect. With no ventilation and very little airspace underneath the plastic, it gets very hot under there. As long as it's sunny and 65°F or hotter outside, within 24 to 48 hours the greenhouse effect you create under that piece of plastic will kill most weeds that are left over after harvesting a crop.

The two main challenges to solarization are weed species that are hard to kill quickly and the weather not cooperating. Hopefully, you don't have a lot of deeply rooted perennial weeds in your field, but if you do, they may be able to survive 24 to 48 hours under clear plastic. Deep-rooted perennial weeds are a challenge for any kind of agriculture. They may need to be removed by hand and/or monitored diligently so they don't go to seed. Eventually, the population will decrease.

Weather can complicate solarization. The main rule of thumb is this: the warmer and sunnier, the better. So, when you don't have those conditions, it's a problem. For example, if you had a bed of one crop that was harvested one day and you wanted to put another crop in a day or two later, but it ended up raining both days, obviously solarization is not going to work. Sunshine is required.

Solarization can work really well and quickly to kill many weed species; however, some species (especially deeply rooted perennial weeds and grasses) may be able to survive even lengthy solarization — like the grass shown in this picture creeping in from outside the hoophouse.
CREDIT: ANDREW MEFFERD

Soil Life and Solarization

The mini greenhouse effect that you create by trapping the heat from the sun in a very small area between a piece of plastic and the soil (and without any ventilation) can easily get temperatures above 100°F on a sunny day. Some people wonder what kind of effect these temperatures have on soil life; if it gets hot enough to kill the weeds, what would it do to the soil life?

Soil is *not* a very good conductor of heat; if you put a thermometer probe through the clear plastic and into the soil on a patch of ground that is being solarized, you will see that the soil temperature returns to normal very quickly as you push it deeper into the soil. Also, keep in mind that we are required to heat compost piles to between 131–170°F for it to be considered compost for organic standards. So even though that's hot enough for killing weeds, solarization is not a sterilization procedure any more than the composting process is.

Here's single-bed solarization at work on Bryan O'Hara's Tobacco Road Farm in Connecticut. Under warm, sunny conditions, solarization can eliminate the weeds remaining after a crop has been harvested in as little as 24 hours.
CREDIT: ANDREW MEFFERD

Solarization can work really well, as long as there's enough heat and sun. Solarization killed a lot of the weeds in the bed shown here. However, you can see by the green still showing that some of the weeds were just fine and grew quite happily — like they were in a little greenhouse. This photo was taken during a very mild summer, so solarization didn't produce very high temperatures under the plastic. In this case, purslane and a few other weeds were able to survive. CREDIT: ANDREW MEFFERD

Tarp Timing

Tarping can be done any time of the year. But solarization only works during warm parts of the year, when temperatures are above 65°F and it is sunny. Depending on which weeds are present (and how many), solarization can be effective in as little as 24 hours. But, even though solarization can work in as little as 24 hours when there is relatively low weed pressure and hot sunny weather, more weeds and more perennials and/or cooler weather will make it take longer or not work at all. We had about a quarter of an acre of tarps down this past summer for weeks straight, but we had an unusually cool summer; some purslane and other weeds simply grew all summer long under that piece of plastic.

Above: *This photo, from Spring Forth Farm in North Carolina, shows a cover crop before tarping.*
Credit: Spring Forth Farm

Right: *This photo shows the same field after tarping. Active soil life has consumed the entirety of the cover crop to leave bare soil. Factors that affect how quickly plant residue breaks down under a tarp include the amount of residue, how small the pieces are, temperature, moisture, and how active the soil life is.*
Credit: Spring Forth Farm

Although tarps won't produce a lot of effect until the warmer part of the year, that doesn't mean that winter isn't a great time to put down tarps! In fact, it's one of my favorite times to put them down. I'm usually less busy in the wintertime so might as well do the jobs that can be done then. Winter tarping can also be a good solution in areas that get a lot of snow, like I do in Maine. If I can get tarps down right before it snows, the snow cover will keep them on all winter long and I don't have to worry about them blowing away. And then as soon as temperatures warm up in the springtime, the soil life will become more active and start doing its job down there in the dark.

For tarping, the longer you can keep them on the better. As a grower, you should take this to mean that if you have plans to grow in a given area, it's to your advantage to get them on that ground as soon

In this photo from my visit to Seeds of Solidarity Farm, this tarp went down on bolting greens crops about a month before this picture was taken. The crop residue has the consistency of dry hay and can be raked off the beds for immediate planting, or the tarps can be left on as a placeholder.
CREDIT: ANDREW MEFFERD

as possible. Unless you don't want the organic matter under your tarps to break down, the longer you leave them on, the more complete the weed kill and organic matter breakdown will be. That's why I say, even if you're not going to grow on a spot for another year, throw the tarps down now or as soon as you get the chance, and they can be working for you while you do other stuff, and then you just have to pull them back to get planting.

Decomposition Rates

If you want a completely clean field, there are things you can do to speed up the decomposition process. One of the most notable ways to do this is simply to use no-till methods for several seasons to encourage as much life as possible in your soil. This is a recurring theme: there are a lot of things about no-till systems that will work better and better the longer the systems are used. In this case, the trend you should see is that the digestion of organic matter by the soil speeds up as your soil becomes more alive.

I have heard from many growers that, as their soil became healthier, they had increased numbers and diversity of microorganisms and their organic matter broke down more quickly. This can be a good thing or a bad thing depending on whether you want your organic matter to stick around as mulch or if you want it to disappear! However, there are a lot of things you can do to affect how long the organic matter sticks around on the surface or gets digested into the soil.

Though nothing is going to incorporate organic matter into the soil as quickly as a plow or rototiller, the microorganisms will do it for you, albeit more slowly. The smaller microorganisms will work on breaking down plant matter between the surface of the soil and the tarp, and larger organisms, like worms, will pull pieces of plant matter into the soil to eat them. So all that organic matter will eventually get worked into your soil.

Moisture

One of the many factors that will affect whether your plant matter stays on the soil as a mulch or breaks down into soil organic matter is *moisture.* Plant matter breaks down faster when it is wet. When dead plant matter is dry, it tends to stick around a lot longer. Because I live in a humid climate in Maine, I don't see this effect a whole lot in the field (the way I might if I lived in a place where it was dry over the summer). But I do notice this in a hoophouse where, of course, it never rains.

One year I had some plants in the hoophouse, and I covered them with a tarp to kill them and hopefully break them down to clear out the bed for the next crop. To my dismay, when I pulled the tarp back, dead plants were still there on the soil surface, making it impossible to plant. The problem was that the bed was dry when I put the tarp down; so, although the plants died, it was too dry under the plastic for them to break down. So I had dead dry plants on the surface instead of the bare soil I wanted.

One of the benefits of not rototilling is being able to leave irrigation systems down between bed flips. Luckily, in this case, I had left my drip irrigation under the plastic, so I turned it on and left the tarps down for a few more weeks. The addition of moisture helped break those plants down. If you're growing somewhere that doesn't have irrigation, you can plan accordingly. For example, you can put a tarp down at the end of your rainy season, when the ground is naturally wet before going into the dryer, hotter summer season. Or, if you live in a dry area, do an irrigation before you put the tarps on, and that will help the plant matter break down.

Temperature

Another factor that will affect how quickly tarping works is *temperature.* Warmer temperatures will speed up the occultation process, because if you put the tarp down on living plants, the hotter they are, the faster they want to grow, and they exhaust their reserves more quickly. This is one reason why it is most common to see the black side of a black/white silage tarp facing up; it absorbs more heat and speeds up the process.

If you live in a really hot climate, you could put the white side up to have the opposite effect and *moderate* the temperature. Whatever you use for tarping, make sure it either has a black side or is completely opaque. I say this because some tarps (for example, tarps that are just white or blue, without a light-blocking black layer) let enough light through for plants to grow. They can actually create a pretty good greenhouse environment for weeds to keep growing underneath your tarp, which is obviously not what you want.

Another factor that will affect how quickly plant matter breaks down under a tarp is how fine the plant matter is. Intact plants will break down more slowly than plant pieces. That's why *flail mowers* have become popular as a way to mow when you need crop residue to break down quickly. They are preferable to rotary mowers like bush hogs because the action of the flails tends to "chop and drop" organic matter in place, rather than blowing it around and shooting it out to the side in bigger pieces, like rotary mowers tend to do.

Area of Application

Occultation and solarization can be used either on big blocks or on the scale of individual beds. However, in either scenario, if you lay the tarps down in strips, and the strips are surrounded by grass, it will tend to infiltrate the edges and grow into the beds. So, you can do strips, but there will be less grass infiltration if you do strips surrounded by a mulch, like wood chips, instead of grass. Otherwise, you'll constantly be beating the grass back from the edges of your strips.

It's a good idea to mow, roll, or otherwise flatten a cover crop before tarping. This picture is of a tarp that went down on sorghum Sudan grass without mowing. This turned out to be a mistake. Water couldn't drain off the tarp, and the uneven tarp surface increased the chance of the tarp blowing off.
CREDIT: SPRING FORTH FARM

IT'S ALL ABOUT THE MULCHES

IF I ONLY HAD A MINUTE and had to summarize for somebody how you prepare land for planting without tilling, my answer would be: it's all about the mulches. When I say mulch, I'm referring to anything inert that separates the soil from the sky: tarps, thick layers of compost, shredded or chipped wood, plastic film, landscape fabric, and more. All these materials can be thought of as "mulch" because they are all something that you put on top of the soil to *keep something in, to keep something out*, or *both*. For example, you might put a layer of mulch on the soil to keep perennial weeds from coming up. Or, you might put a mulch on the soil to protect it from heavy rains that could cause erosion.

Most of the successful no-till methods use mulches in one form or another to first clear the space for planting instead of tillage. And then we need a way to suppress weeds without deep disturbance to the soil during the growing season. Some of the methods, like using tine weeders, stirrup hoes, or other types of cultivators, have already been discussed. They can all be used to cultivate weeds — even in a no-till system — without deep soil disturbance.

An alternate solution is to rely on mulches to suppress weeds, with even less soil disturbance. I say alternate, because when deciding between cultivation or mulching, it is usually an either/or situation.

Since chunky mulches tend to clog up cultivators, the decision is usually between planning to suppress weeds with a mulch or to take them out with cultivation. The exception would be something like compost, which could be cultivated if weeds germinated in or came up through it. Some of the methods used in no-till function both as soil prep *and* weed suppression. For example, a deep layer of compost gets the bed ready to plant at the beginning of the season and suppresses weeds throughout the growing season.

Most growers use more than one method of no-till at the same time, so it is not so much about choosing a one-and-only method as it is about knowing which method to use when. In a single season, you might use several methods: the roller/crimper method for a big block of crops like squash, sunflowers, or hemp; occultation to prepare a smaller block of mixed vegetable and flower crops; and solarization to quickly kill the weeds between crops.

With mulches, you might even use two at the same time. For example, you could prep a field in fall for the following spring with a deep layer of compost as mulch; then you could put a tarp over the compost to keep weeds from coming up over the winter. If you live in an area with a lot of winter precipitation, the tarp over the compost would also keep the field from getting too wet over the winter. Or, if you live in a dry area, you might put a layer of wood chip mulch on top of the compost mulch to conserve moisture.

Mulches of all types will help retain moisture, but they can also help when fields are too wet. An impermeable tarp applied before big rains to decrease the amount of precipitation that soaks into a field is yet another trick that can be used to get planting earlier with no-till. Regardless of whether the precipitation to be avoided is a winter's worth of snow accumulation or a large amount of rain during a rainy season, a tarp can be left on a field to keep rain off and then peeled back for planting as needed. If you want water infiltration while your mulch is on, use landscape fabric or another permeable mulch.

Mulches can be split into two broad categories: those designed *not* to break down so that they can be removed later (non-decomposing), and those left in place to decompose and eventually become a part of the soil (organic). The biggest difference between using a non-decomposing mulch vs. an organic mulch is the longevity of the material. There are some very different dynamics with a mulch that doesn't decompose vs. one that does, so think about which features you need for your farm. Many farms will use organic and synthetic mulches at different points in the season, to meet different goals.

Non-Decomposing Mulches

The subject of tarping in relation to solarization and occultation has already been discussed in "Tarping," above. But tarps are a type of mulch because they can also be employed mainly to keep things *out* (rain, weed seeds) or *in* (moisture, micro livestock, germinating weed seeds).

One of the paradigms that no-till turns on its head is the use of plastic mulches on the field during the growing season. In tillage agriculture, it has become very common to put down black plastic film mulch or synthetic landscape fabric under many transplanted crops, especially heat-loving ones like tomatoes, peppers, and eggplant. These synthetic or plastic mulches are left on the field during the growing season. They are usually very thin and only last a season, meaning that they need to be applied in the spring and then removed in the fall after the crop is out.

Removing this type of mulch is a very dirty job, and it generates a lot of plastic waste. It also usually leaves bits of plastic or microplastics in the field, depending on how degraded the plastic got during the growing season. In no-till, the plastic tarps tend to come *off* during the growing season; it's much more common for plastic mulch to be applied and then removed *before* the cash crop goes in the ground (as in occultation and solarization). And it's often organic mulches that are used on the ground to suppress weeds during the growing season.

It's not that plastic film mulch can't be used in no-till, but I haven't seen a lot of it. It's easier to put plastic film mulch down on smooth

It's difficult to mow weeds growing over the edge of plastic mulches like this landscape fabric because a mower would chew up the material. CREDIT: DAN PRATT

Grass growing over the edges of mulch can drop a lot of seeds in a very short time. CREDIT: DAN PRATT

beds that have plenty of loose soil to bury the edges, i.e., tilled soil. In my experience, it's less common to see single-use plastic film mulch in any form in no-till. And when plastic mulch is used with a growing crop, most often I've seen landscape fabric used with holes punched in it for the crop. This type of mulch does have the advantage of being reusable.

Although tarps are in the category of "non-decomposing," it is wise to keep in mind that they do *degrade*. We are all aware of how much damage plastic is doing to the world, but, at the same time, it's almost impossible to farm using zero plastic. That's why I think it's a good trend that in no-till oftentimes the plastic mulches come off the field during the cash crop; the growing season is usually when plastic gets damaged and destined for recycling or, more likely, the trash. Although plastic particles may last forever, even the toughest plastic tarp will eventually break down in the sun, get holes in it, and need to be discarded. One of the positive things about no-till is that it tends to use reusable tarps as we strive to find something better than plastic.

PLASTICS ON THE FARM

Unfortunately, even organic farms are as addicted to plastic and fossil fuels as the rest of the modern world. Possibly even more so, when you think about the outsized significance of greenhouse film, landscape fabric, tarps, flame weeders, and black plastic film mulch to veg and flower farming. When farming regeneratively, it's important to recognize where our methods are not perfectly sustainable so we can improve them.

Dare I speculate that the cheapness and availability of plastic tarps are one of the things that have fueled the current popularity of no-till? It's my suspicion that our golden age of soil science, combined with the ease of throwing a tarp down for field preparation, may have given growers both the knowledge and the tools to make a transition. But now that the methods are proven, many growers are trying to think of viable alternatives to the plastic that has become indispensable on our farms.

There is so much plastic already in the world. We could coast for a while on reused silage tarps, landscape fabric, old greenhouse covers, and other repurposed materials. However, we're going to have to think of better alternatives at some point so we can shed our reliance on plastic.

Canvas tarps represent one tantalizing possibility. They can act as a replacement for plastic tarps, but they have some notable drawbacks. For one thing, canvas tarps are much more expensive than plastic tarps, so that will be a barrier to adoption. I've also heard mixed results about canvas tarps; some growers think they work just as well as plastic tarps, others point out that they become very heavy when waterlogged, and that they may break down more quickly than plastic tarps.

I try not to rely on techno-fixes that don't exist yet for solving problems, though it is my hope as we pull away from plastic as a culture, that biologically based polymers will fill the needs that plastic is currently filling. It may sound overly optimistic to think about some bio-based material replacing a durable plastic tarp. However, we know that even the sturdiest plastic cover will eventually develop holes and break down. So a material doesn't have to last forever to work as a tarp; all we have to do is find a happy medium between plastic and a durable biopolymer that lasts long enough.

When you consider the amount of money that has been put into the research and development of plastic over decades, you can imagine a similar investment and urgency to get off of plastic would yield a similar array of materials that would — hopefully — be just as useful as plastic. Ideally, they could be composted or otherwise disposed of in an ecologically sound manner at the end of the product's life cycle. Imagine a bio-based tarp that you throw on the compost pile when it reaches the end of its useful life. I'll be first in line to buy one.

Even if weeds and grass are not completely gone when the tarp comes off, a top dressing of compost will serve as mulch to suppress weeds so the crop can get established without competition, as with these rosemary plants on Bare Mountain Farm.
Credit: Andrew Mefferd

Tarps as Placeholders

In a perfect world, a cover crop is the perfect placeholder because, in addition to suppressing weeds like a tarp, it also adds organic matter and feeds soil life with root exudates (for the discussion about cover crops, see "Mulches Grown in Place," below). However, there's not always time to get a cover crop in, and sometimes you need a quick placeholder. For example, in beds that have been harvested late in the fall or early in the winter, too late to seed a cover crop, a tarp is a great option to suppress winter weeds, hold in fertility, and help break down organic matter left over from the previous crop.

Though some systems depend on being completely weed-free to eliminate competition with direct-seeded or other small plants, other systems tolerate some weed pressure as it does not interfere much with large transplants that can quickly shade out weeds, as with these zinnias. As you can see, they are very productive, despite some weeds and grass that want to encroach on the bed. In this system, as long as weeds don't go to seed, they are not a problem because they can get knocked back by tarping between crops. CREDIT: ANDREW MEFFERD

The strip of landscape fabric on this fallow field is being used to maintain the field edge by smothering encroaching grass.
CREDIT: ANDREW MEFFERD

Tarps don't have to be an all-or-nothing affair. One valuable way to use the placeholder ability of tarps is to gradually bring a field into production. For example, let's say you tarped half an acre with 30 beds in it, and you only need a few beds at the beginning of the season. You can fold the tarp back bed by bed as you need them, prepping individual beds and leaving the rest covered and weed-free until you can get to them. It's very refreshing to have a field that can sit, and not feel like the clock is ticking on weeds sprouting. You can just leave the tarp on and wait until it's time to plant.

At Four Winds Farm in New York State, they use comfrey planted on the edges of the field to keep grass from encroaching. The perennial plant's root system is so dense that grass has a hard time growing through it to get into the field. CREDIT: ANDREW MEFFERD

Another way to maintain edges is with hay bales on field borders. Of course, they don't last forever, but they will keep grass from encroaching and can be used as mulch or composted when they break down. CREDIT: ANDREW MEFFERD

Edges don't always have to be perfect; they just have to keep weeds back long enough for the cash crop to get established. For example, as this photo from Bare Mountain Farm in Oregon shows, repeated tarping can keep knocking the grass pathways back from the beds. Though weeds may keep trying to encroach, tarping between crops ensures that the flowers are big enough to shade out the weeds by the time they try to grow into the bed. CREDIT: ANDREW MEFFERD

Keeping Tarps Down

One of the biggest challenges to using any kind of tarping material is keeping it down. I learned this the hard way as my first few tarping attempts were blown off the field. I learned to put more weight on tarps than I thought they needed to keep them down. We've used pallets, empty plastic bulb crates, cinderblocks, T-posts, rocks, and pieces of wood. Basically, anything that is heavy enough to hold the tarp down and doesn't have any sharp edges or points that would pierce the tarp can work. We have a lot of T-posts on the farm, and when they're not in use for trellising, they make great weights for the edges; they are so long and heavy, they can keep the wind from getting under tarp edges.

Though weighing down the edges might be the most important part of keeping tarps down, don't forget about the middle of the tarp! Even if you weigh down the edges really well, strong winds can get

In this photo from Bare Mountain Farm, you can see how they use plenty of sandbags to keep the strips of landscape fabric from blowing off. This landscape fabric is slightly wider than the beds they are covering to allow for some crowning of the tarp by the residue from the previous crop. This is important to keep in mind if you plan on tarping beds with a lot of residue on them.
CREDIT: ANDREW MEFFERD

the center of a tarp flapping so much that it can shake the weights off the edges. Plus, the mechanical action of flapping will also cause your tarps to break down more quickly. Put something in the middle of your tarps so they don't flap.

Applied Organic Mulches

Organic mulches are convenient because you don't have to pick them back up, and they break down and add to the soil over time. They are great for adding to soil organic matter, but not so great for long-term weed control, since one of the things we like about them (they break down) also means they won't control weeds indefinitely. So, think about your farming goals and mulch accordingly.

You can use more types of organic mulches in no-till systems than you can in most tilled agriculture because you can use some highly

We dealt with a weed problem in the pathways of this hoophouse by mulching the pathways with cardboard covered with hay to help keep the cardboard down.
CREDIT: ANDREW MEFFERD

carbonaceous materials that would tie up nitrogen if they were plowed into the soil fresh. Many of the materials that are good for mulching between crops are also good at forming a durable pathway because they don't break down quickly. It's also more practical to have persistent mulch as pathways if you're not plowing up and rearranging the field every year. Mulching pathways is a lot of work to do, only to have to do it again every year. Though we know some people who remake beds and pathways every year, it's not for everyone.

Traditionally, a big limitation to mulching in growing areas has been the need to cultivate; most cultivators need nearly perfectly trash-free conditions or they get clogged up by the materials that people use as mulch. In systems that don't cultivate, you can use as much mulch as you want on the bed tops because you don't have to worry about cultivation tools.

Growing beds surrounded by sod has some advantages; the sod can serve as perennial pathways that don't get muddy. But this approach also has disadvantages; as soon as tarps or cover crops are off, grass will start encroaching into the beds. Also, it's difficult to keep grass from going to seed on the edges of tarps without mowing the tarps themselves, because seed heads can grow out over the tarps. Some systems can withstand the constant pressure from the edges by knocking the grass back with repeated tarping.
Credit: Dan Pratt

Organic Mulch Materials

The most commonly used organic mulch materials that are applied to the soil to suppress weeds during the growing season are compost, straw, leaves, wood chips, and cardboard. But almost any organic matter that is free of weed seeds and can be relatively cheaply had can be used. Whether you live in a rural area or an urban area, think about what organic matter byproducts may be generated in your region. For example, is anyone near you generating bark, wood chips, nut shells, spent brewer's grains, cardboard, sawdust, or anything else that could be used to block weeds? Some of these materials may even be free depending on how badly the people who are generating them need to get rid of them.

You can use wood chips more liberally as a mulch in no-till because they sit on the soil surface and get a chance to break down partially before being incorporated into the soil via layering. In fact, tillage growers usually steer clear of wood chips altogether because, as a highly carbonaceous material, when they get tilled into the soil before they have broken down, they tie up nitrogen as the soil life digests them. Sitting on top of the soil as mulch gives them a chance to weather before becoming a part of the soil by layering.

This photo from Alnarp's Farm in Sweden shows how you can turn a tedious job into a fun one by having a tape removal party after stockpiling enough cardboard for a new field. Under organic rules, mulching is only permissible with cardboard with black ink, since there is a risk of chemicals in other color inks.
CREDIT: MARIE-CLAIRE FELLER

This photo from Seeds of Solidarity Farm and Education Center shows how cardboard can be used as a mulch for transplants, like these tomatoes in a greenhouse. The cardboard is applied over the compost or other bed materials. Then transplants can be planted through holes in the cardboard, and the cardboard can be covered with straw to help keep it down. Credit: Andrew Mefferd

HOW MUCH MULCH TO APPLY?

How much of any given mulch to apply can vary a lot, depending on the goals and history of a farm and properties of the mulch(es) to be used. Mulches can be divided into two categories: those that can do double duty as both planting medium and weed-suppressing mulch and those that simply block weeds from coming up.

The only commonly used mulch that functions both as a planting medium and a mulch is compost. When getting started with no-till, it is very common for growers to build their beds by applying somewhere in the 4-inch range of compost. Compared to the amount of compost many growers apply at field scale, that is a lot of compost. However, most growers look at an application like that as a one- or two-time thing to get beds established, to 1) provide enough loose compost for transplanting or direct seeding, and 2) suppress weeds from coming up from the native soil.

These two goals taken together add up to why many growers start off with so much compost: the need for friable soil and something to control a heavy weed load. Hopefully, once you've got friable soil established and the weed seed bank is going down over time, you can taper off with the compost. Many growers end up applying a dusting on a yearly basis, or whenever they transition a bed to a new crop, depending on how much they need the fertility, organic matter, and loose planting surface provided by compost. For growers who need to keep applying compost after they've got their fertility high enough, in order to keep raising soil organic matter or for a loose planting surface, they can use compost made out of leaves or other materials that are high in carbon and lower in nitrogen than traditional compost. In this way, they can keep applying organic matter without overloading their soil with too much fertility.

Though 4 inches of compost is a common starting point for growers establishing no-till beds for the first time, growers have used as little as 2 inches to get beds started where weed pressure isn't as high and where a lot of loose soil isn't needed. For example, growers transitioning to no-till on fields that don't have a lot of weed pressure and/or are direct seeding (so don't need as much loose soil) might opt for less compost.

On the other hand, growers who anticipate a lot of weed pressure might double up on mulches. Four inches of compost will smother a lot of weeds, but those who anticipate high pressure of perennial or other persistent weeds, or are trying to build beds on sod that isn't entirely dead, might add a layer of cardboard or paper under the compost for a little more weed-suppressing power.

When we get into the loose organic mulches, sometimes thickness is determined by the material itself. For example, unrolling a round bale of hay tends to provide a few inches of hay mulch in a long strip as wide as the round bale is tall. That is enough to smother most weeds, with thickness depending on how the hay was baled. (See the photos of us preparing our hemp field for this in action in the last chapter.) It's a great way to quickly mulch a bed or a pathway, since the bales unroll as quickly as you can move them. Even square hay or straw bales tend to separate into flakes that are a 2–4 inches deep. If you're shredding hay or straw, either by hand or with a bale grinder, at least 2–4 inches will probably be needed for weed control unless weed pressure is very light.

The other organic mulch materials, like wood chips, sawdust, nut shells — or really any other weed-free, affordable, nontoxic organic matter — can be applied in varying thicknesses based on how you are using them. For example, 2–4 inches of any of these materials can be applied in paths to make a durable, weed-suppressing pathway. The same materials can be applied more in the 1–2 inch range to the top of a bed to provide a weed-suppressing mulch; sometimes they are added to bed tops in even lighter amounts, sometimes barely covering the soil, where weed pressure isn't high, to suppress some weeds and lessen soil splash and crusting from hard rains where that is a problem.

How much mulch you apply really depends on what material you are using and what your goals are. You might even apply the same mulch in different thicknesses in different areas; for example, you might put 4 inches of wood chips in walkways for a durable path, and an inch or 2 of wood chips on the bed top for weed suppression. Just be wary of any mulches that might have weed seeds in them. That could include compost, hay, straw, and many other organic materials.

Like so many things in agriculture, know what the source of your material is and what the risks are. If you're not sure and you get offered a good deal on some hay that got rained on in the field, you could take it and leave some out in the rain (or water it) to see if anything sprouts. If weed seeds do germinate, you can leave it out in the rain until no more weeds are sprouting. Otherwise, materials that have an unknown load of weed seeds can be more trouble than they're worth.

Organic Mulch Pros

In addition to there being a wide variety of materials that can be used, another advantage of organic mulches is that, depending on the material, they can be applied quickly. For example, it is possible to apply mulch and plant on the same day. An extreme example of this would be Ricky Baruc's Insta Gardens, where he puts down cardboard over sod and then puts compost on top of the cardboard and plants into that. By the time the roots of the crop have grown down to the cardboard, the sod is dead.

Another good thing about organic mulches is that the weed suppression can be excellent as long as the mulch is thick enough. And organic mulches contribute to soil organic matter as they break down. Regardless of whether used in beds, in pathways, or both, mulches can keep produce and farmers less muddy, and reduce the amount of soil splash that gets on crops. This can help speed up postharvest handling and washing and can also help reduce diseases. This is particularly important for tomatoes and other crops that are vulnerable to diseases transmitted from the soil to the leaves by splashing water from rain or irrigation.

Organic Mulch Cons

Although some organic mulches can be applied quickly (cardboard, compost, and landscape fabric are examples), others can be time-consuming to apply. For example, mulching by hand with straw bales, and applying some other types of mulches by hand can be time-consuming. Any possible way to mechanize or speed up the process should be considered ahead of time. For example, in most manure spreaders with rear beaters, straw bales can be put into the bed of the spreader, and then the strings can be cut and the spreader turned on, and the rear beaters will shred and spread the bales while the moving floor of the manure spreader keeps the bales moving into the beaters.

Another way to quickly apply straw or hay is to buy round bales and simply cut the strings and roll them out on the area to be covered. If you've never done it before, round bales unroll like jellyrolls; one or two people can usually unroll one. See the photos of us planting our hemp with this process in the last chapter, "Case Study: Growing Hemp."

One implement that can speed up and increase the precision of spreading thin layers of compost is a drop spreader like this one from BCS America.

CREDIT: PHOTO BY BCS AMERICA

They cool the soil, for better or worse

Though there are dark-colored organic mulches, many of the commonly available ones are light-colored, so they will keep the soil cool — the opposite effect of using black plastic mulch. This can be an advantage as long as you want cool soil; for example, light-colored mulch can help grow cool-season crops in hot weather. A light-colored mulch might help broccoli or other brassicas grow deeper into the summer. But a light-colored mulch is not usually the best choice for tomatoes because they are a heat-loving crop.

Lots of mulch may attract pests

There are a number of ways that mulch may attract various creatures that are undesirable for your crops. For one thing, it may give pest species more hiding places. A good example is voles. Voles are notorious farm pests (especially in the wintertime), and a tarp or a deep layer of straw mulch may provide a hiding place for voles to hang out in between the times they are eating your crops. So, be aware that if you already have vole pressure, deep mulches may make it worse — at first.

This is one of the many areas where you may notice no-till methods working better over time. Pest explosions are frequently followed by an explosion of pest *predators*. So, even though at first vole populations may explode, if you can stick with the no-till methods, their predators will probably catch up with them. Over time, the populations should even out, and you can go back to whatever you were doing to control voles in the first place.

Slugs are also notable mulch lovers, and they don't need a very deep layer for mulch to be attractive to them. Their soft bodies don't like to be out in the sun, so they will frequently crawl under mulch during the heat of the day only to reemerge to continue feeding on your crops at night. Knowing this tendency of theirs, one way to catch slugs is to leave pieces of cardboard or wood in the growing area, and you will likely find slugs hanging out on the underside during the day. You can flip the wood over and slide the slugs into some soapy water or use some other method of getting rid of them.

Just as with the voles, an initial slug explosion may be followed by an explosion in populations of ground beetles, snakes, and other slug predators; hopefully, over time they will balance out with the slug population. However, if the slugs get out of control, you can try using Sluggo, an organically approved slug bait. Slugs are attracted to it, and then they die when they eat it. It's one way to control slugs until slug predators can do it for you.

This is another area where you need to take into account your climate and all the other relevant details of your farm. After I wrote *The Organic No-till Farming Revolution,* I heard from one guy who said he couldn't believe that I was recommending that much mulch and that he had terrible slug problems using deep mulch. Well, it turned out he was in a very humid semitropical environment that was basically a slug breeding ground; in extremely warm, wet climates, deep mulch may not be a good option due to slugs. But it can work well in a lot of other climates.

In this sequence, we are reclaiming a field we have not grown on in a few years from sod.

Our first step was to put a tarp down the previous fall, so by the time we pulled it back six months or so later the following May, the sod was pretty well killed except for some perennial weeds. But we know that the rhizomatous grasses will creep in from the edges, so we put cardboard down at the edges of the beds to keep the grass from encroaching. CREDIT: BY MARY HALEY, MXH MARKETING

After broadforking, we added four inches of compost to build the beds. This gave us two things: loose material with plant nutrients and biology to plant into, and a means to suppress any weeds that might come up from below. When we get farther away from the compost pile, we use a tractor to dump compost into wheelbarrows and then use the wheelbarrows to distribute the compost to the beds. Because we are not planting this whole field at once, we are leaving most of the tarp on and only peeling it back bed by bed as we build them so the rest of the field stays tarped and doesn't start growing weeds. CREDIT: MARY HALEY, MXH MARKETING

Once we have compost on the beds, we use the back of a rake to flatten it out. The back of the rake is less aggressive than the tines and better for flattening. Don't waste too much time on this step, it doesn't have to be perfect, just enough to evenly distribute however much compost you are using.

With a bed of nice loose compost, transplanting or direct seeding can go quickly. We like to have one person go ahead and throw transplants at the correct spacing on the bed top, and have other people come behind to tuck the transplants in. It's a lot faster to have one person pop all the transplants out of the flats and drop them where they need to be planted than it is for the same person to switch back and forth between popping plants and transplanting. CREDIT: BY MARY HALEY, MXH MARKETING

Transplanting goes really quicky when you can just poke transplants into a loose layer of compost with your hands.
CREDIT: BY MARY HALEY, MXH MARKETING

MULCH GROWN IN PLACE

THERE ARE TWO WAYS you can get mulch on a field: you can either *apply* it where it needs to be, or you can *grow* it where it needs to be. In this section, we are going to talk about the latter.

When you think about how bulky, heavy, and awkward transporting organic matter for mulch (straw, wood chips, compost, etc.) is, it's really a brilliant idea to just grow the mulch where you need it to be. That's also why methods based on mulch grown in place are the most likely candidates for being scaled up. Even though you might need an enormous amount of mulch to cover a multi-acre area, planting that area to a cover crop only requires a relatively modest amount of cover crop seed.

However, getting the crop established and then *crimped* so you can plant your crop are the tricky bits that require planning, the right timing, and the right crops in order for it

This is a large roller/crimper that can be mounted on a tractor 3-point hitch.
CREDIT: ANDREW MEFFERD

Side view of the roller/ crimper. You can see how, when the fins are in contact with the ground, they put concentrated pressure on the stems they are rolling over, crimping and terminating them (as long as you get your timing right).
CREDIT: ANDREW MEFFERD

all to work. I became acquainted with the efficiencies and challenges of the roller/crimper method when I worked on the research farm at Virginia Tech. It was the method we were researching to compare organic no-till systems to tilled organic systems to see how they stacked up against each other in yield, quality, and profitability.

The Roller/Crimper Method

The whole idea behind the roller/crimper method is that you grow a very dense cover crop that, when you roll it down flat, will form a dense mat that weeds can't grow up through. The sequence goes like this:

1) establish a cover crop, 2) roll and crimp the cover crop down flat, and then 3) plant through the cover crop. Weeds are suppressed by the rolled killed cover crop. One notable advantage of this method is that it can add a lot of organic matter to a field without you having to handle or apply it yourself. So you get all the benefits of a cover crop plus the benefits of an organic mulch.

Though it may be possible to eradicate weeds completely (or almost) on a fairly small intensively managed plot, it is much harder to get larger acreages sto be completely weed-free. This is why when we talk about scaling up, it's important to have a weed-suppression strategy, like a mulch, unless you have a way to cultivate those weeds that grow in your no-till system.

This is a transplanter that we used at the Virginia Tech research farm to quickly transplant seedlings into high-residue beds that had roller/crimped cover crops as mulch. The main modifications were the large straight colters in front of the soil-loosening shanks and the furrow openers. Without these colters, which cut through the flattened cover crop mulch like a giant pizza cutter, the residue would clog up the transplanter very quickly. Many other off-the-shelf planters could be modified similarly; this one was otherwise a standard transplanter. In one pass, it can 1) cut a slit and loosen soil for transplants, 2) water them in, 3) drop longer-term pelleted fertilizer from the metal boxes, 4) lay drip tape, and then 5) transplant. CREDIT: ANDREW MEFFERD

This direct seeder has been retrofitted to work in mulch grown in place that has been rolled and crimped. The white drums at the back sit on top of the seeding mechanism and can hold large-seeded crops like corn or beans. The large black colters in the front cut through the cover crop so that the seeders don't get hung up on the crop residue.
CREDIT: ANDREW MEFFERD

In order to act like a mulch, the cover crop not only needs to be dead but also needs to be *flat* on the bed. In most cases, it is not sufficient to simply roll a cover crop down with a smooth roller. Even though a smooth roller may flatten the vegetation, many cover crop species (especially the vigorous ones we like to use in no-till, like rye and vetch) are not reliably killed just by flattening them. If you do just roll them down without crimping, there's a good chance that they will pop back up or experience significant regrowth. This is why the roller/ crimper was invented.

A roller/crimper is simply a round drum with fins welded to the outside so that everywhere a fin presses into the stems of your cover crop, it *crimps* them; where the fin presses into the stem it makes a kink and disrupts the flow of water through the plant's stem. This

then prevents moisture from making its way from the roots to the top of the plant, effectively killing the plant even when it is in a vigorous stage of growth. This is why timing is so important because the prime time to terminate most cover crops is when they are in flower, just before making seed and growing vigorously. The idea is that when you crimp a cover at the pollen stage (flowering), it is more susceptible to being killed mechanically.

If you try to crimp too early, before flowering, the cover crop may actually be growing too vigorously, and it will have a good chance of regrowing. If you crimp too long after pollination (once viable seed

You can see how much residue is on the beds from the rolled/crimped cover crop; the retrofitted transplanter can plant right through it due to the colters that cut a slit in front of the transplanter shoes.
CREDIT: ANDREW MEFFERD

IS THERE A MINIMUM SCALE WHERE MULCH GROWN IN PLACE MAKES SENSE?

I used to think that cover crops grown in place, to be terminated for mulch with a roller/crimper, was only an appropriate technique for larger farms. This is because the research farm where I first encountered the method was set up with tractors and expensive pieces of equipment that could plant large areas quickly. Then I started visiting with farmers who had come up with ways to grow and crimp a cover crop in place with hand tools, on a scale as small as a single bed at a time.

So now what I would say is that if you want to do no-till on a large scale, the roller/crimper method is the easiest to scale up quickly and extensively. But thanks to the ingenuity of growers who figured out how to do it on smaller farms, it can be cheap and easy to try on a small scale. This diversification of methods is useful for growers both big and small; smaller growers may want to intersperse beds with roller/crimped mulch in with different methods for different crops. And for larger growers, being able to try a few beds or blocks of roller/crimped cover crops is important to make sure the method and timing work for the desired crops before scaling up.

has formed in the cover crop), you are too late because even if you kill the cover crop, you just planted your own weeds with the viable seeds that the cover crop will drop. Know how long the period between pollination and viable seed formation is for whatever cover crop you plan on roller/crimping because that is the sweet spot for mechanical termination and will tell you how big of a hurry you have to be in to terminate it once you see that pollen!

Roller/Crimper Cons

Cover crops grown in place can be a great and time-efficient method of weed control; however, there are drawbacks, as with any method.

Cost of Equipment

The cost of equipment for a mulch-grown-in-place no-till system on a large scale is higher than the cost to start a no-till operation on an acre with some tarps and some hand tools. At the very least, you will need to own or rent 1) a tractor, 2) a no-till drill or some other planter to plant the cover crop, 3) a roller/crimper to terminate the cover crop, and 4) some kind of no-till seed planter or transplanter to make slits in and plant through the mulch.

However, the proportion of the farm dictates the proportion of the equipment; if you are running a multi-acre farm, buying all that equipment will be a boon in terms of time saved and the ability to scale up. I learned the roller/crimper system on the research farm at Virginia Tech, where we were using equipment with the ability to plant acres of produce or flowers in a day, if not hours. Since starting my research on no-till, I have been impressed with the innovations farmers have devised to use mulch grown in place on smaller acreages that don't justify expenditures on big equipment.

Smaller-scale growers have been able to use the system by seeding cover crops into bare soil created by occultation and/or putting some compost on the bed top to create a seedbed with enough loose soil to bury a seed for germination. After the cover crop grows, some farmers employ small roller/crimpers that can be propelled by a *walking tractor.* I have also seen growers using a T-post or a piece of angle iron attached to a bottom of a 2 × 4 (really, anything that can be stepped on to crimp the cover crop will work) to terminate a cover crop.

If you want to try this, attach a rope to both ends of a 2 × 4 with a T-post or angle iron secured to the underside; or you can just use a T-post by itself to get a good crimp. Adding the 2 × 4 to the T-post or angle iron simply makes a larger, more comfortable step for the farmers to put their weight on with every crimp. Two farmers can walk down a bed, each one picking up the crimping implement by the rope on the end, moving it a step forward and stepping on the crimper to drive it into the crop and crimp the stems. It's the farmer version

This is one way to crimp a cover crop without a roller/crimper. There is a rope tied to each end of the T-post, and each farmer uses the rope on their end to lift the T-post along with their foot as they take a step forward and bring it down on the next section of cover crop, flattening and crimping it at the same time. Some growers prefer to attach a 2×4 to the top of the T-post to provide a larger surface to step on. CREDIT: PHOTO COURTESY OF JEN ARON OF BLUE RAVEN FARM IN OREGON

of a three-legged race. If you find it hard to picture this, take a look online. There are a bunch of videos of farmers doing this.

Mulch grown in place can be tried on a small scale on most farms with little expenditure; many farms already have a direct seeder to plant the cover crop in the first place (like an Earthway, Jang, or Planet Jr.) and a T-post or 2 × 4 lying around with which to crimp the crop.

Transplants and some large seeds (which tend to be stronger and able to germinate in rougher environments) can be put in with a trowel or other simple tool that will cut through the cover crop mulch as you go. This is an ultra-low-cost method that can be great on small acreages if you only need to do a few beds at a time. If you're spending a lot of time on this, it will give you the opportunity to contemplate whether it's worth spending the money on equipment to expedite the process.

Requires Planning

As already mentioned, there is only one crimping window for most crops: it comes after pollen drop but before viable seeds have formed. This is the period of time when many crops are most vulnerable to death by crimping but have not yet made viable seeds — which means you have to plan backward and have planted the cover crop with enough time to be at the right stage for crimping when it's time to plant your cash crop. In many cases, for example with late spring or early summer cash crop planting windows, that planning may extend to the previous season because you need to have planted a cover crop the previous fall or winter. For example, with rye or vetch, for them to overwinter and be at the right stage for termination when your crop needs to go in, they need to be planted the previous fall so they will survive the winter and make the rest of their growth before crimping the following spring. In all but the warmest climates that can germinate rye, vetch, or other cover crops over the winter, overwintered fall planting is the only way to get enough biomass for late spring/early summer crimping.

It might seem like, for later-season crops (those that you aren't in a hurry to plant as soon as you can in the spring), you could just terminate the cover crop anytime in the spring and wait until it's time to plant. But the clock starts ticking on two important factors as soon as you crimp a cover crop. For one thing, as soon as a cover crop is dead, it starts decomposing. So, the clock is ticking on how long the mulch will last as soon as the crop is crimped. It is ideal to plant into a cover crop mulch shortly after it is crimped so as to take advantage of its full weed-stopping power.

The second thing to consider is that, once the cover crop has been crimped, weeds may start germinating. So, if you terminated a cover crop and left it for an extended period of time, you would probably come back to find weeds growing up through the mulch. This is especially true with rye, because it has allelopathic weed-seed germination-suppressing qualities from biochemicals it secretes; these start to wane as soon as the rye is dead.

Having studied the roller/crimper method at Virginia Tech, we were all ready to start our farm using this method. But then we realized that we would have plantings every single week of the growing season, including lots of fine-seeded crops, like salad mix. And we saw that the crimper method was not conducive to either a schedule that had lots of different planting windows or the need to have small-seeded crops germinate in the field.

Because we hadn't yet been introduced to all the other methods in this book, we ended up just starting our farm as a tillage-based farm. Happily, though, we returned to no-till when we started reading about all the other farmer-developed methods. So, beware: the roller/ crimper method can be really effective, but not for every single crop or every single situation.

There is a good list of the cover crops that have been studied for no-till and their characteristics. It is overly long to include here, but you can search for "Choosing the Best Cover Crops for Your Organic No-Till Vegetable System" at rodaleinstitute.org.

Not Never-Till

When originally conceived of, the crimper method was not *never-till*. It was acknowledged that it was not a complete solution to the problems that came with using cover crops as mulch. A gradual increase in weed pressure was to be expected, and fields would need to be plowed every few years, when weed pressure got too high. Though this fulfilled the aim to till less, this need for periodic tilling has been a dealbreaker for some who want to never till again.

However, much depends on how you manage the cover crop. It is possible to use the crimper method without ever tilling. But for the method to truly be never-till, you have to have a way of establishing a cover crop in untilled soil, like using a no-till drill, for example. It also helps to have fairly weed-free soil to start with, like a well-maintained vegetable field rather than, say, a hayfield since it's hard to establish a cover crop in sod.

If someone did want to manage the roller/crimper system without any tillage at all, it could be done — as long as there is a way to establish a cover crop and stay on top of whatever weeds do come up. It's important that they don't get out of hand. In a perfect world, no weeds would come up under our crimped mulch cover crop; in reality, though, there are likely to be weeds. And though it is possible to get a field that is almost weed-free, you won't get there unless you have a *plan* for dealing with the weeds that do come. Since almost no type of cultivation is possible on beds with as much residue on them as in the roller/crimper system, your plan may involve spot-treating weeds that do come by pulling, hoeing, or some other method of eradication.

As already discussed, for the roller/crimper method to work, cover crop termination timing needs to be carefully matched to crop planting timing. A wide variety of cover crops can be used as roller/crimper mulch cover crops to facilitate different timings. You just have to figure out a planting timing and cover crop (or combination of species) that can be terminated just before your cash crop needs to go in.

Another method that can help you manage planting windows is tarping after crimping. If you need to crimp a cover crop before it's ready to die and want to avoid any regrowth, putting a tarp over it should help *really* kill it. The combination of rolling, crimping, and tarping should be enough to kill even the heartiest cover crops. It's not ideal, though, in that it adds the extra steps of putting on and removing the tarps. Plus, you need a couple weeks for the tarps to sit on the field and do their job (more or less, depending on how vigorous the cover crop is). However, if you have a planting window that is really hard to hit given your combination of available cover crop species and weather, you can probably make it work with crimping and tarping.

WHY RYE AND VETCH ARE DOMINANT COVER CROPS IN NORTH AMERICA

Though there are many different species of plants that can be used as cover crops, there are two that are probably planted more than everything else combined. For a few reasons, *rye* and *vetch* are very commonly used in North America, in no-till and tillage situations alike. First of all, we should clarify that we are talking about winter rye (*Secale cereale*) and hairy vetch (*Vicia villosa*); there are other species of vetch and rye that may have different applications but are not as universally adaptable as winter rye and hairy vetch.

Rye and vetch are so popular, on their own as well as in combination, because they are very strong growers and will put on a lot growth quickly, which is important whether you want to smother weeds or add organic matter. Vetch is a legume that can fix nitrogen for your soil. Winter rye has an allelopathic effect, meaning that during growth it secretes chemicals that suppress the germination of weed seeds. This effect dissipates over the weeks after the rye dies, so it's a good idea to wait a few weeks after the rye is dead before direct seeding anything into rye stubble.

Both rye and vetch can survive the winter or be winter-killed, depending on whether or not you need the cover crop to start growing again when it warms up. This involves some planning. Rye and vetch that are close to maturity when freezing conditions set in are likely to die in the cold weather, so you can plan on planting earlier in the season if you want them to winter-kill. Rye and vetch planted a few weeks before the first frost date in the fall are likely to overwinter as small plants, and then start growing again in the spring. Beyond that, local conditions vary so much it's important to find out the common planting dates in your area; these two are so widely grown there should be data available for your region.

When terminating rye or vetch yourself with a roller/crimper, you need to kill them when the crop is making pollen but before viable seed has formed. Pollen is more visible in rye, so a better indicator for the vetch would be when you see some seedpods starting to form on the vetch flowers. Note that vetch has complex flowers that form, open, and pollinate from the top down. Since the pollen is harder to see, you know this has taken place when you see the first few flowers start to drop and seedpods start to form in their place. It's important to roller/crimp it before viable seed forms in those top pods.

You can look up how long it takes viable seed to form for many different species, but like most other biological processes, the plants develop faster the warmer it is and will take longer to mature seed the colder it is. Anyway, it's best not to rely on numbers on a chart when it comes to calculating your crimping window. An unusually cool or warm season can really throw those dates off. When you see flowers, you know there will be pollen shortly. When you see pollen, you know there will be seedpods. You can go out there and pop some seedpods open to make sure the seed isn't mature. Use your eyes and know that when you see the progression of flowers, pollen, and then seedpods, you know there's seed developing and you need to roller/crimp it soon before that seed matures.

For all these reasons, rye and vetch are a good starting point for learning about cover cropping. However, there are many other species that can grow in different conditions. For example, oats are a common cover crop that reliably winter-kill; Austrian winter peas provide an alternative to vetch that can also overwinter; and in very hot climates, cover crops like sorghum Sudan grass and sunn hemp provide very fast growth with a lot of biomass.

Sometimes, when you plant a diverse mix of seeds, a few species will take over. This is a good indication that the species that took over are suited to your climate and season and are worth trying again; cover crop species that don't thrive may simply be planted out of season for your area, so study the crop descriptions and perhaps try them at a different time of the year if it seems like they should grow in your climate.

If you're not used to growing cover crops, rye and vetch are widely adapted and relatively easy to grow, so once you've gotten those down, find a chart or, even better, talk with other growers in your area to find out which other cover crops grow well in your area and can be added to the mix.

Requires Very Dense Plantings

You need a really dense stand of a cover crop for this method to work. If the cover crop isn't planted heavily enough, if it wasn't fertilized enough, or if something disrupts germination and growth, you may get unacceptable levels of weed growth. The problem could be something that affects the whole field, like a seeding rate that is too low or not enough fertilizer. Don't go light on cover crop seed. Trying to use a crimped layer of mulch that is not thick enough can be a disaster because weeds may grow up through the mulch unimpeded.

Various other factors can also result in spotty weed control. For example, if a heavy rain washes cover crop seed out of part of a field, or if part of the field did not get enough fertilizer, the part of the field that suffered may result in localized weed problems. For the crimper method to work, you've got to take your cover crop as seriously as your cash crop. The cover crop cannot be an afterthought. Don't use old seed. Use enough fertility. And get the cover crop planted on time. Failing to pay attention to any of these factors can result in not having enough cover crop mulch to effectively suppress weeds.

Difficulty with Direct Seeded Crops

You will need to keep in mind that the roller/crimper method doesn't create the fine seedbed that is needed for germinating a lot of the smaller-seeded crops. This is not to say that you can't direct seed into a roller/crimped cover crop; thousands of acres of corn are planted into untilled soil every year. But corn is a big seed. If you tried to seed, say, salad mix with its tiny little seeds into a roller/crimped bed, it may not work out very well.

Because most of a bed may be covered in mulch, access to the soil tends to be in small strips that have lots of residual roots in them. Small seeds need good soil-to-seed contact, and crimping does not usually create the environment that small-seeded crops like to germinate in. Some growers have had luck with germinating seeds using *zonal tillage* in crimped fields. This means leaving most of the soil untilled, except for a small strip where the seeds will go.

Direct seeding anything into a roller/crimped bed will require making a slit or a channel through the mulch that the direct seeded crop can be planted into. On a field scale, we used a shank with a colter in front of it to both cut through the mulch and loosen the soil just enough to receive seeds. Still, the combination of lots of roots in the soil and only a very small slit through the mulch may not result in good germination with small-seeded crops. This is one reason why the roller/crimper method has predominantly been used with transplants and large-seeded crops so far. If you want to use it with a small-seeded crop, do a test on a small area before planting a whole field to make sure your seeds can germinate.

Winter-Killed Cover Mulch Crops

Another possibility for terminating a cover mulch crop is to deliberately let it *winter kill*. For example, garlic could be planted into an established oat crop in October, and the oats will continue to grow in the fall before the garlic puts on any top growth. By the time the garlic sprouts in the spring, the oats will have been killed by the winter

This is a crimper designed to roll a bed at a time.
CREDIT: DAN PRATT

temperatures, and they won't compete with the garlic crop. Of course, for this method to work, you need to be in a climate that has winters cold enough to kill the oats.

Where I live in Maine, we get enough snow (about 100 inches a year) that the snow usually does the job of crimping for us when it falls on and flattens the dead oats. If you live in an area where it is cold

Above: *Side view of crimping.* CREDIT: DAN PRATT

Right: *Tarping can allow for the establishment of a cover crop for rolling/crimping or a cash crop, though grass will start growing in from the edges as soon as the tarp is off.* CREDIT: DAN PRATT

enough to kill the oats but don't get enough snow to flatten them, you may need to roll the oats down after they've been killed but before the garlic comes up through them. Otherwise, if the oat residue is still standing when the garlic comes up the following spring, it will not form a weed suppressing mulch mat.

Tomatoes planted into compost in a crimped cover crop.
CREDIT: DAN PRATT

Adding Compost and Amendments

Some people prefer to do their bed prep ahead of time by putting down fertilizer and/or compost when they plant the cover crop; this simplifies and speeds up planting the cash crop when the time comes. For this method, you just have to remember to apply enough fertility for *both* the cover crop and the cash crop to follow. On the other hand, some people prefer to add fertilizer and compost over the top of the bed before planting the cash crop. This can be done with a manure spreader, a drop spreader, or by hand.

Another alternative for applying fertility for the cash crop, one that many growers use with larger transplants like tomatoes, is to dig a hole with a shovel in the roller/crimped cover crop and put a shovel full of fertilizer in the hole where the plant goes.

A drop spreader is the most efficient way to put an even layer of compost on a bed.
CREDIT: DAN PRATT

One way to prep beds is to sprinkle the necessary amendments on the bed tops, dump compost in regular piles, and then rake it out for an even seeding or transplanting surface.
CREDIT: HILLVIEW FARMS

So, there are many considerations when it comes to deciding when and how to apply fertility to crimped beds. One rule of thumb would be to fertilize entire beds for smaller transplants, like broccoli, and dig holes for larger transplants, like tomatoes. One of the benefits of following this rule is that you can provide long-season plants (like tomatoes) with the greater amount of fertility that they need; they benefit from the larger hole with compost. Most shorter-season crops (like broccoli) and the smaller transplants require a higher planting density; it would be much too time-consuming to dig a hole and fertilize every single one for densely transplanted crops.

Planting through a Cover Crop Mulch

On a small scale, plants can be put in quickly by hand through a cover crop mulch using a trowel or a shovel, depending on the size of the transplant and the hole. Another option is to use a bulb transplanter to take a chunk of soil out and drop a transplant in. This can be efficient when one person goes ahead making holes and another comes behind to do the transplanting.

Sometimes, however, with a dense cover crop with a lot of roots that haven't had any time to break down, all the interwoven roots can make for tight soil that can be hard to break through. If this is the case, running a shank through the soil can be done to speed up the transplanting process. If you have a tractor, you can pull a shank preceded by colters through the channel where you want to plant the crops. You'd be surprised how much simply running a shank through can loosen up some soil and make it fairly easy to get transplants in the ground. Instead of having to make a hole for each plant, if channels have already been cut for the plants, then planters can come along behind and quickly place the plants in the channels, instead of having to stop and make a hole for each one. The next level of efficiency in speeding up planting would be to use a mechanical transplanter. There are some made specifically for no-till that have extra colters and openers in front to keep the residue from building up where the transplanter opens a furrow for plants to go into the ground. Some transplanters can be retrofitted with colters to work in a high-residue environment (as shown in the photos near the beginning of this chapter).

Which Crops Work Best in the Roller/Crimper System?

We've already discussed the difficulties of getting small-seeded crops to germinate well in crimped cover crops, but crops that can be transplanted are a natural fit for mulch grown in place. Because they can go through the most vulnerable stages of growth (germination and small-seedling stages) in a propagation greenhouse, transplants are a way to get around germinating small-seeded crops in a roller/crimped

Because mulch that is grown in place and rolled/crimped may be scaled up more quickly than some of the other methods, it is a good choice for crops that take up a lot of space and are often grown in blocks, like tomatoes, pumpkins, winter squash, gourds, hemp, or these sunflowers. As the mulch breaks down over time, the ability of plants like hemp, cucurbits, and sunflowers to shade out any weeds that make it through the deteriorating mulch is a bonus.
CREDIT: ANDREW MEFFERD

field — where they might not have enough seed-to-soil contact to germinate well.

An added benefit of large plants that form a canopy is that, later on in the season, when the cover crop mulch is starting to break down and is not suppressing weeds well anymore, larger plants will be starting to close canopy and do their own weed shading and suppression. For this reason, crops like pumpkins, sunflowers, squash, tomatoes, and hemp have worked well in roller/crimper fields; they shade the weeds themselves once the mulch breaks down.

For crops that aren't as competitive with the weeds but still need a long season of weed suppression, there is one work-around developed by Shawn Jadrnicek. He adds chopped

leaves on top of the mulch to increase the amount of time that it will suppress weeds for. If you don't have leaves, you could augment the crimped cover crop with any other organic mulch material that could go down on top of the crimped mulch and still be planted through.

In summary, the roller/crimper method may not work for every vegetable or flower crop, especially those with small seeds that need to be direct seeded. On the other hand, it is probably the method that can be scaled up the most quickly, by using no-till drilling or otherwise planting a cover crop on a more extensive acreage — provided that the system works with the crops that you want to grow. As with anything in farming, if you're not sure it's going to work, test it on a smaller area before going big.

Alnarp's Farm in Sweden uses cardboard covered by wood chips to keep grass from encroaching on the perimeter around their beds. They build their beds with a layer of salvaged cardboard on the ground, covered by six inches of compost to form the beds, and wood chips in the pathways.
CREDIT: ALNARP'S FARM

After prepping the entire hoophouse, we tarped the whole area to help smother any weeds that might come up, and peeled the tarp back as we needed to plant additional beds. CREDIT: ANDREW MEFFERD

GETTING STARTED AND CROPPING STRATEGIES

I F YOU'RE READING THIS BOOK thinking about starting a farm, my best advice is to get good at growing whatever you want to grow before you start your business. One great way to do this is to apprentice or otherwise work on someone else's farm. Even if the work isn't on a farm in the region where you want to grow or they don't use the exact techniques you want to use, anything you grow will add to your understanding as a grower.

Like anything in farming, I would recommend trying something on a small scale before betting the whole farm on it. Whether it be changing tillage techniques or growing a new crop, things don't always go the way we think they will. The closer the dress rehearsal can be to the final, the more effective it will be. In the past, people may have felt like they needed to rent or buy land and a tractor to start a farm. And this is where lowering the barriers to farming comes in. One of the things that makes me the most excited about no-till is that people can start tarping and growing on almost any available land, whether before starting farming full-time or before having an actual farm property.

As a matter of principle, start preparing your land as soon as you get access to it, ideally at least the season before you want to grow on it. As soon as you think you want to farm a particular piece of ground, you should be cover cropping, or tarping or solarizing or employing whatever techniques you're going to use to start getting the ground ready. Remember, though, that while no-till techniques can offer great things in improving soil, most of them are not as fast as plowing. Plan accordingly.

Composting in Place

There are two applications for compost that have outsized importance in no-till, and one of them is to compost plant residues right where they are. This is a slow method to transition beds from one crop to the next, but it does save the time and labor of removing the remains of a crop to a compost pile. Which is a great time-saver as long as you are not in a hurry to crop an area again soon after the first crop is done.

The basic method is the same as occultation, though you will need a tarp that is slightly larger than the area to be tarped if there is a lot of residue, as with high biomass crops like corn, sunflowers, sorghum Sudan grass, etc. This method can come in handy at the end of the season if you don't have enough time to plant a cover crop, or any other time when one crop is done and you don't need the space right

If not many weeds, like an already established field, a field that has been tarped for a month or more, or a roof, solarize or mechanically weed first to get rid of any light weeds.

Beginning of the season — what state is your field in?

If heavy weeds or sod, tarp your field (occultation) or plant a cover crop to smother what's already growing there. Or, if you don't have enough time for a round of tarping, you could plow one last time to get rid of whatever is growing there.

Bed prep — are you planting seeds or transplants?

For seeds, if the soil isn't loose enough to run a seeder through, make a seedbed by either adding enough compost to bury the seeds or use a tilther, rake, rotary harrow, or other implement to make enough loose soil at the surface.

For transplants, if the soil isn't loose enough to transplant into, add a few inches of compost, or make holes with a trowel or shovel, or run a shank down the row to open up a furrow.

How to handle weeds?

away. You don't want it to sit there growing weeds, so you can simply put a tarp over the area and walk away until you need it.

If the tarp is there long enough and your soil life is active enough, there may not be any residue left by the time you pick the tarp back up. And remember that, as your soil becomes more alive, your residues will break down even faster. You will learn how much time you need for your soil to digest crop residue, or conversely how much mulch you need to put down to keep it there for the duration of a crop.

Compost as Mulch

As already mentioned, in addition to being used for fertility and soil building, compost is also sometimes used as a mulch in no-till. However, if you use a lot of compost as mulch, year after year, there is a definite possibility of overloading the soil with fertility. It may sound like there are worse problems to have, but overfertilization is not something we want. The runoff from overfertilized fields can pollute waterways, and overfertilized crops are more prone to a variety of pest and disease problems. Plus, it's just a waste of fertility. This is

This chart summarizes the turning points in a season where a decision needs to be made between one management decision or another.

CREDIT: ANDREW MEFFERD

Mulch to suppress weeds. Weeds that do come up through mulch probably have to be pulled by hand, since most mulches clog up mechanical cultivation.

With no mulch, have a plan for mechanical cultivation when weeds are as small as possible to minimize soil disturbance and time spent weeding.

At the end of the crop...

If there is not much crop residue (for example, after a cleanly harvested bed of carrots with few weeds), go back to the bed prep step before planting the next crop.

If there is a lot of residue from the previous crop, remove and compost elsewhere or compost in place before going back to the bed prep step.

At the end of the season, either...

Tarp the field to act as a placeholder to keep weeds from growing and retain nutrients from leaching. If you need winter precipitation to infiltrate the soil, put on a porous tarp like landscape fabric. If your winters are too wet, put on a solid tarp like a silage tarp to prevent the soil from getting soggy over the winter.

OR

Plant a cover crop. Take the residue from the last crop off the field if there's enough that you can't plant through it. Plant with a no-till drill or the same method as you use for direct seeding.

Get ready for another season and go back to the beginning!

When you have the time to prep beds for planting the previous fall, it can really streamline spring planting. The beds shown here were prepared before being tarped in the fall. In this case, landscape fabric with pre-cut holes for plants went on the beds because they are planted in a field known to have high weed pressure. In the spring, all that had to be done before planting was for the tarp to be peeled back (from left to right in this case) and the beds were ready, irrigation and all underneath. CREDIT: ANDREW MEFFERD

one reason why many growers who want to use a lot of compost on a regular basis as mulch will make or buy a high-carbon compost. Growers who want to speed up soil-building by applying higher volumes of compost could be in danger of overfertilizing a field if applying a lot of hot (high-nitrogen) compost year after year.

So, if you're standing looking at a field that you want to plant to a vegetable or flower crop, here are the questions you need to be asking yourself: is the field sod or has it recently been used for veg or flower production? Looking at the "Getting Started" chart, go to

field prep for sod versus plowed fields. Are the crops that you want to grow space-intensive or space-extensive? Other factors you need to take into account are: climate, soils, crops to be grown, labor versus mechanization, and field crops or patches vs. beds. You can use the chart seen here to help you decide on a procedure for all of these issues.

Ways to Improve Land and Exhaust the Weed Seed Bank

The fastest way to improve land is to take it out of cash crop production for a short while and focus on things that will make the land healthier, so it has more to give to your crops. It is also possible to address other growing issues while improving the land. For example, if you wanted to grow more of your own fertility, you might incorporate legumes into your cover crop mix for the nitrogen that they will fix and add to the soil (don't forget to inoculate the seed with nitrogen-fixing bacteria!).

Or, if your biggest problem is annual weeds, you could do several cycles of irrigation and tarping until no more weeds are germinating. Or try one of the other options described below for improving the land. Of course, being able to take land out of cash crop production in order to improve it with a cover crop requires having enough land that you don't need to be cash cropping 100% of it all the time. Because land will improve faster when you are not cash cropping it and you can just focus on improving the land, it's ideal to have at least a little more land than you need for cash crops so you can always be improving some of it.

The Not-Quite-Bare Fallow

Not-quite-bare fallow is a modification of an idea I heard from some of the best farmers I know, Anne and Eric Nordell of Beech Grove Farm, in Pennsylvania. Many organic systems recommend taking 20% or so of your land out of cash crop production every year for cover cropping. The idea behind the *bare fallow* is that the more land you take out of production through cover cropping in any given year, the more quickly your land will improve. In the past, Anne and Eric have written about bare fallowing up to 50% of their ground to improve the soil and work down the weed seed bank. I have been to their farm, and the results are impressive. And though not no-till, the principle is the same whether you're tilling or not; the more land you put into cover crops or other rotations designed specifically to improve the land instead of cash crops, the faster your land will improve. You can read their own description of the method in the November, 2017 issue of *Growing for Market* magazine: "Whoa-till: Minimum-depth Tillage for the Dryland Market Garden,"(growingformarket.com).

The basic idea behind their bare fallow is that they take ground that is weedy, and just let the weeds grow, and then do multiple rounds of shallow plowing. The plowing just skims off the surface of the soil, killing the weeds. Several rounds of this, along with cover cropping, helps improve the soil and work the weed seed bank down more quickly than is possible with a cash crop in the ground to work around. The no-till version of this would be to let the weeds grow, and then do multiple rounds of tarping instead of plowing to kill them.

Being Weed-Free Is Not a Fairy Tale

Having weed-free fields isn't magic; though it's rare enough that people may think it's an impossible dream, I've seen a few fields that were well managed enough to have hardly any weeds, like the Nordells' fields mentioned in the last paragraph. I think a lot of farmers who plan on cultivating just accept that there are going to be weeds ... because they plan on cultivating. But if you go into the project with the goal of exhausting the weed seed bank (see "Stale Seedbedding" section, above) and stay on top of weeds, it's more likely that you'll work your weed seed bank down to a minimal level, and mostly just have to deal with weeds that blow in or are otherwise brought into the field.

CROPS TO FOCUS ON

YOU CAN GROW ANY CROP you want in a no-till system, but particularly with newer growers in mind, I would advise focusing on crops that there is a ready market for. Especially if you are just starting out, focus on high-value crops like salad mix; flowers; head lettuce; bunched greens; and vining, fruiting crops (tomatoes, cucumbers, and peppers, especially). Even if you grow many other crops, these staple crops keep a lot of market farms in business because people want to buy them year-round, and they are lucrative enough to make a profit on. This is in contrast with, say sweet corn, which is popular with customers but takes up a lot of space to grow, and corn is notoriously difficult to make a profit on. Until your farm business is in the black, spend most of your time on the most profitable crops that pay the bills.

ONE-CUT SALAD MIX

Though most crops can plug right into no-till, there are a few that make especially good use of the system. One of those is lettuce bred to be grown to maturity for processing into salad mix — known as "one-cut lettuce." This is in contrast to the way salad mix has been grown traditionally, like mesclun mix for example, which is produced from the cut leaves of immature plants. This is why the traditional method is sometimes called "baby leaf" whether the species grown is lettuce or some other species of greens.

People want to have salads all year-round, so lettuce and salad mix are disproportionately important crops to many market farms. A perennially popular item with shoppers, lettuces are great because they have the potential for year-round sales and can be grown year-round in most locations. This has made lettuce a staple of market farm sales.

When one-cut lettuces came out in 2012 or so, we quickly saw the advantage for our farm and switched from growing salad mix by direct sowing and harvesting baby leaves to transplanting and harvesting mature heads and then processing them into salad mix. Even though we weren't no-till at the time we switched, we could see that increased yield and quality would make one-cut more profitable for many farms than growing baby leaf salad mix, whether no-till or not.

"One-cut varieties" refers to types of lettuce that are bred so that all the leaves join at a single point at the base of the plant, similar to how pac choi normally grows — with all its leaves anchored to the growing point at the base. One cut just above the growing point will release all the leaves — thus the name, one-cut lettuce. It's hard to recommend specific varieties, because the varieties frequently change and they are released in series that should cover most of the colors and forms that growers would need. We have found the Salanova and Eazyleaf series do the best for us.

Once they had the idea, breeders worked for years to develop varieties that would allow them to be grown and harvested with the efficiencies of head lettuce, and easily processed into salad mix. When all the leaves connect at the same point at the base, whole heads can be harvested in the field and brought back to the packing shed where the base can be cut off to release the leaves.

Now, of course, it's not that nobody was harvesting whole heads for salad mix before 2012; people have made salad mix for a long time by chopping up leaves from mature heads of lettuce or other greens. The difference with these one-cut varieties is that they are bred to have all small (baby-sized) leaves, unlike normal head lettuces that tend to have big leaves on the outer edges and tiny leaves in the center.

The biggest advantages of growing lettuce to maturity for salad mix are higher quality, higher yield, longer shelf life, and better eating quality. Let's look at each of those advantages individually.

Growing lettuce as heads yields a higher-quality salad mix because you don't end up with the ragged brown edges that come from chopping head lettuce small enough for salad mix. This is important because all those cut edges and little bits of leaf that are created when head lettuce is chopped are where decay starts. This is one of the reasons why one-cut lettuce usually has a better shelf life for salad mix than either baby leaf or head lettuce chopped into salad mix.

Using mature lettuce to produce salad mix is higher yielding because when you direct seed salad mix, some of those days to maturity are spent simply with the seed sitting in the ground, waiting to germinate. During that time, the ground is uncovered, and, because no root exudates are going into the soil, all the sunlight that is hitting that good soil is wasted and not growing anything on top of the soil or feeding anything in the soil.

Let's look at the typical days to maturity for salad mix versus one-cut lettuce. Seed catalogs will tell you that many baby leaf varieties clock in at about a month to maturity, and many of the one-cut lettuces clock in at closer to two months. So you would think it would be quicker to produce baby leaf than one-cut lettuces.

However, one detail that growers can take advantage of is that half of the roughly 60 days to maturity for head lettuces can be spent in a cell in a propagation greenhouse. This means that (if you grow transplants) you can cut the number of days one-cut lettuce needs in the growing bed by half. So now you've got your one-cut taking as much time as baby leaf.

The other dynamic that helps one-cut lettuces be more profitable is that they are in the neighborhood of 40% higher yielding than most baby leaf lettuce. So, by growing one-cut lettuces for salad mix, you can maintain roughly the same cropping schedule — in other words, harvesting almost every 30 days out of any given bed space — with up to 40% higher yield. And, in addition to that, because no-till beds don't have to be out of production while they are tilled and reshaped, the harvest-and-replant-in-the-same-day method possible with no-till can speed up the cropping of one-cut even more.

In addition to increasing yield, using transplants can also make salad mix production more reliable. When direct seeding baby leaf, sometimes a planting may not come up perfectly. This might be due to insufficient rainfall, poor soil-to-seed contact, thermal dormancy in the summer; sometimes, it's just a mystery why a direct seeded planting doesn't germinate well. Skips in a planting or incomplete germination for whatever reason can result in your coming up short on salad mix that week and/or having to replant and hope for better luck.

Transplanting lettuce means you'll never have this problem. Germination percentage should be very high when plants can be sprouted in the controlled environment of a propagation greenhouse. And if there are germination problems, they can be remedied (i.e., replanted in the prop house) before outdoor growing space is wasted on poorly germinated stands.

Another advantage to planting transplants of lettuce instead of direct seeding baby salad mix is that it keeps living roots in the soil and the leaves will protect the soil above ground for more of the year. When you direct seed, in the time between when the seed is put in the ground and when it sprouts, there is no root activity, and then there's a week or two when the roots are so small they're not providing much in the way of exudates to the soil life. In this way, transplanting salad mix is better at meeting the goal of keeping the soil life fed and the soil surface protected.

Yet another advantage of harvesting mature lettuce for salad mix versus true baby leaf is in shelf life. You may grow a salad mix that tastes wonderful, but if it doesn't make it to your customer's plate in good shape, nobody can enjoy it. So, even though shelf life is not the sexiest thing to grow for, everybody likes it. Mature leaves have a longer shelf life than baby leaves, so one-cut lettuces harvested for salad mix will last longer than true baby leaf. One-cut lettuces will also outlast regular leaf lettuces that are chopped for salad mix because the one-cut lettuce only has the one cut at its base versus the multiple cuts needed to make mature head lettuce into baby leaf size pieces. As I already mentioned, each cut edge and, even worse, the little bits that come from chopping up leaf lettuce into salad-mix-size pieces are what start to decay first. No one wants to smell that rotting smell when they open their bag of salad mix.

Even if it isn't actually rotten, increased shelf life will keep salad mixes more appealing for a longer time. This is one of the happy examples where the features the farmer wants coincide with what the eater wants. Moreover, mature lettuce is also nicer to eat; it has more flavor and crunch because the ribs of the lettuce leaves are more fully developed.

In summary, one-cut lettuce grown for salad mix may be able to increase the number of successions and the yield a grower can get out of any given space. Plus, it can increase shelf life, flavor, and texture, all at the same time. When we switched from baby leaf to one-cut lettuce, our retail and wholesale customers noticed; they commented on how nice our salad mix was. That was really gratifying because it's not like we told them: "Oh, we started doing our salad a different way." They just noticed that something was different that set it apart from other salad mixes they had access to.

Profitability of One-Cut Salad Mixes

Besides the fact that I like the flavor and longer storage of mature one-cut head lettuce over baby lettuce leaves, they also yield higher; transplanted head lettuce should yield up to 40% higher than direct seeded baby leaf after the same amount of growing time, with somewhere in the range of 270#/1,000 ft^2 for mature lettuce vs. 190#/1,000 ft^2 for baby leaf. So yield and quality are higher for one-cut lettuces, but the seed is more expensive. At a 1 seed/

foot planting rate for one-cut lettuce vs. 60 seeds/foot for direct seeded baby lettuce, seed cost is approximately one-third more: about $190 to plant 1,000 ft^2 of one-cut lettuce and $145 to plant 1,000 ft^2 of baby salad mix. It does end up costing more to plant the same area with one-cut lettuce, even though one-cut requires fewer seeds.

So, there is a little more seed cost upfront for a one-cut lettuce mix, but the other costs are buried in the cost of owning and running a propagation greenhouse. If you start seedlings in a greenhouse, you're probably already running it for other crops, so it's likely not a big deal to grow some lettuce seedlings. You can successfully direct seed one-cut lettuce and get rid of the cost of propagating the seedlings, but then you tie up your growing space as the seed sits in the ground for a few days germinating and then growing very slowly for a couple weeks. So, if you don't have a propagation greenhouse, you can still grow one-cut lettuces by direct seeding them; they're just going to take up growing space for longer than if they were transplanted.

Figure out what the economics are and how they apply to your farm. Or, maybe you just don't like taking care of baby plants and want to direct seed everything! Suit yourself. Farming involves personal style and preference. Luckily, there are as many ways to grow healthy crops as there are farmers. Say you're getting $10/pound for salad mix. The breakdown of gross profit on the different ways to grow salad mix goes something like this: $2,700/1,000 ft^2 of one-cut salad mix per month-long crop cycle; $1,900/1,000 ft^2 direct seeded baby lettuce mix per month-long crop cycle; or $1,350/1,000 ft^2 for direct seeded one-cut salad mix (because direct seeded one-cut takes twice as long as transplanted one-cut or direct seeded baby mix).

So, transplanted one-cut lettuce is both the most expensive to plant and the most remunerative to harvest of the no-till salad cropping cycles. Even with an additional $45 of seed cost for transplanted one-cut lettuce mix, transplanted one-cut still comes out $755 ahead of direct seeded baby mix per cropping cycle, and I would be surprised if the propagation costs for 1,000 lettuce seedlings came anywhere close to $755. Though I imagine if you're already direct seeding all your lettuce, and don't want to start, either of the direct seeding options (one-cut for heads or closely seeded for baby mix) are profitable enough to be worthwhile.

With transplants, you can spend this whole germination/baby stage with plants at high density in a propagation greenhouse, and you are also putting a plant in the ground with a root system that is big enough to immediately grow into the living support system for the soil life you're encouraging. Another advantage of transplanting is that germination conditions can be better controlled than with direct seeding.

GOING FROM CROP TO CROP

FOR LONG-SEASON CROPS like tomatoes, cucumbers, peppers, and eggplant, there will not be as much interaction with the soil as with quick crops like salad mix. Due to the long-season nature of fruiting crops, you may only have to prep the bed once in the spring, and then you can let the crop grow all season long. The next time you need to do anything with the soil is when you take the crop out and transition to a cover crop or physical cover over the winter.

Quick crops like radishes or salad mix allow for a single growing space to be used multiple times in a year — as many times as you can plant, grow, harvest, and repeat. One-cut lettuce varieties grown for salad mix make particularly good use of the quick, potentially same-day bed turnovers made possible by cutting out the tillage steps in bed prep. It means you can plug lettuce seedlings into a bed, then come back in as little as a month to harvest full-sized heads that are easily processed into a high-flavor, long-shelf-life salad mix. (See sidebar, "One-cut Salad Mix.")

Much of how you transition from one crop to the next will be determined by how quickly you need to move each bed from an old crop to a new crop. If you need a quick transition, it doesn't get much faster than cutting any crop residue and weeds off at the soil line, adding compost and amendments, and replanting. If time doesn't

I grew these potatoes by putting down fertilizer and amendments on the bed first, then the potatoes, and then rolling out a round bale of spoiled hay to cover them. Some plants had trouble emerging from the hay, but once they got growing they grew well, though the tubers harvested were slightly smaller than when grown in the ground.
CREDIT: ANDREW MEFFERD

matter, pull a tarp over the crop residue — weeds and all — and let the microbes do it for you! A crop that doesn't have a whole lot of biomass, like baby greens, for example, will break down under a tarp quickly, whereas a larger crop with more biomass, like brassicas or sunflowers, will take longer.

Dealing with Residue

A faster way to get rid of crop residue on a larger scale would be to use a flail mower. The advantage of using a flail mower vs. a bush hog or other rotary mower is that it chops plant matter more finely than a rotary mower, and it drops whatever has been mowed right where it is, instead of blowing it around. Not shooting pieces of vegetation out the side can be important when you are clearing out beds that are right next to ones that are still in production; you don't want adjacent crops damaged by flying vegetation. Flail mowers are available in many different sizes to fit a wide variety of walking tractors and riding tractors.

If there is not much residue left from the previous crop, it may work to simply flail mow what is left and let the chopped residue land on the bed to either compost in place or sit on the bed like mulch. On the other hand, if there is a lot of residue left from the previous crop, there are pickup flail mowers available, which will mow and chop vegetation and also collect it to be composted elsewhere.

Once the residue from the previous crop is dealt with, it's time to deal with the soil. I used to have the mindset that every time one crop came out, it was time to fertilize and rototill. Now I know that,

A penetrometer is a tool that can help you avoid unnecessary bed prep by telling you how compacted your soil actually is, which is hard to evaluate accurately by sight or by feel. Read more about it in the article by Jen Aron in the February 28, 2022 issue of *Growing for Market* magazine: "The Penetrometer: A Simple Tool to Decrease Tillage, Labor and Improve Soil." (growingformarket. com).

especially if soil is in good tilth to start with, it's not necessary to till between every single crop. In fact, tilling after every single crop is one of those self-perpetuating cycles that was leading to compaction and soil issues in the first place.

Planting

If your tilth and fertility are in good shape, go ahead and transplant your seedlings. If you're like me and were taught to till between every crop, you will have a feeling like you should be doing something else. But it will go away after a while; once you see it work, you will learn to trust the process. If tillage is the only method you've ever known, I just want you to appreciate how streamlined this method is, when compared to a multi-step soil prep between every crop.

For direct seeding with a push seeder, if you have really loose soil and little soil residue, you may be able to run a seeder right in the bed without further prep; the furrow opener on most push seeders can cut a channel through soft soil regardless of whether it's been tilled or not. Or, if you prefer to broadcast seed by hand, you can sprinkle seeds on the bed top, add compost on top, and (ideally) gently roll the seed to promote good seed-to-soil contact and germination. There's more on broadcast seeding by hand in Bryan O'Hara's excellent book, *No-Till Intensive Vegetable Culture.*

If your soil is not loose enough to seed into, you can use a rake, a hoe, a tilther, a rotary harrow set very shallowly, or anything else to rough up the surface of the soil to bury seeds enough for good germination. The general rule is that seeds

When space doesn't have to be left between rows to allow for cultivation, plants like this baby lettuce mix can be planted at a higher density than is normally used. Though I think it's a good idea to use transplanted one-cut lettuces to shorten the time between planting and harvest, if you don't like transplants, one strategy to get more plants into a given space (and thus a higher yield) is to broadcast seed by hand. CREDIT: ANDREW MEFFERD

should be planted at a depth of 2 to 4 times the diameter of the seed; so, with many small-seeded crops, you really don't need a lot of loose soil at the top of the bed for successful germination.

If you do need to loosen the surface of the soil, add any fertilizer or compost beforehand, and your loosening will serve to lightly mix the amendments. If no loosening is necessary but compost or amendments are, they can simply be applied to the top of the bed and smoothed out with a rake or other means.

If your soil is too tight on top for what you want to plant, or if there are layers of compaction in the soil, broadforking is a minimally invasive way to loosen the soil. A broadfork is a hand tool, usually with 5–7 tines, that can be inserted into the soil and gently rocked back and forth to loosen soil in a minimally invasive way. Apply compost and fertilizer before broadforking, and the action of the fork will work the amendments in a little bit.

Flowers in No-Till

Since soil prep isn't different for flowers vs. vegetables, the advice is the same for how to maintain your soil whether you're growing flowers or vegetables. Though of course there are some differences in flower vs. veg cropping. For one thing, many no-till vegetable growers get a very high return with baby greens and similar short-cycle crops that are grown at high density and turned around in quick rotations. In some cases, they can go from transplant to harvest in as little as 30 days, and then the bed can be replanted to another crop later on the same day after it was harvested. There are no flower crops that offer as quick of a turnaround as salad mix or radishes, but what follows here are some other ways flower growers can get the most out of no-till systems.

Forcing Bulbs

There are a lot of different ways of forcing flower bulbs like tulips. Some people dig trenches and grow bulbs closely spaced in the ground; some people grow them in crates filled with compost or in compost on top of a bed.

First of all, a few words about what differentiates *forcing* from letting things happen on nature's own schedule.

> What 'forcing' means is that the bulbs are chilled down artificially and forced to grow out of their normal season — as opposed to planting the bulbs in the fall in field beds and letting them bloom naturally in the spring. Forcing bulbs are prepared by chilling the bulbs for a set period of time (up to 16 weeks) before bringing them into the greenhouse to grow.
>
> Many spring bulbs such as tulips, iris, hyacinths, and daffodils need this cooling period to produce high-quality flowers. It varies per crop, but every type has its minimum requirement. A typical symptom of insufficient chilling is short stems. This is particularly apparent in hyacinths. More chilling = longer stems.

This definition is from the article "How to Force Tulips," from *Growing for Market* magazine. We don't have space for the rest of the details about forcing, since they aren't directly relevant to no-till. But if you want to read the rest of the article, you can do so at growingformarket.com.

Forcing is different from typical growing methods because you aren't relying on the plant's root system to feed the plant — they grow from the stored energy in the bulb. Moreover, flowers that are forced for commercial production are usually pulled up bulb and all; the flowers get smaller in subsequent years, so most growers find it more profitable to pull the flower with the bulb the first year and buy new bulbs for the next year.

Because the root system isn't feeding the plants, bulbs can be crowded to the point where they are touching each other. You don't have to worry that their roots will compete with each other — you can plant them as close as eggs in an egg carton (they look a little like eggs in a giant carton when you're done).

You can use forcing as a strategy for establishing new beds; it is quicker and easier than digging out trenches. A couple of inches of compost can be put down where you want to grow (just enough to keep the bulbs upright); bulbs can be nestled into the compost, and then they can be top-dressed with one to two inches of compost to cover. Then those bulbs are yanked out as they are harvested, and the space can be used for another crop as soon as the bulbs are gone.

It's particularly helpful to have tulips planted in loose soil or compost. Since you are going to harvest them bulb and all, you can just pull the ones you want to harvest and clip the bulb off into the compost. If tulips are opening faster than you can sell them, don't cut the bulb off! You can harvest them while the flower is still in bud with just a little color showing; as long as they still have the bulb attached, they can be held in a cooler for weeks. The cool temperatures will put the brakes on opening, and the flowers can hold for a surprisingly long time as long as the bulb is still feeding the bloom.

Crowding Flowers for Longer Stems and Higher Yields

Many planting patterns are based on what will be used to cultivate them; for example, if cultivating mechanically, a three-row bed would typically have plants spaced far enough apart for cultivators to pass on either side of each crop row on the way down the bed. So, as long as you're suppressing weeds with mulches and not cultivating with anything that requires exact spacing, you can crowd many varieties of flowers tighter than the catalog recommends.

For a lot of plants, spacing them too tightly results in "legginess," or tall, weak, spindly plants. In flowers, this can be used to our advantage to lengthen the stems of many varieties, which will grow longer when they sense that they are crowded. For cut flowers, this can be an advantage because the height of bouquets or other arrangements is limited by whatever the shortest stem is. There are plenty of great flower varieties that just aren't used in the cut flower trade because their stems aren't long enough; varieties that are borderline long enough may be limited to shorter arrangements. Crowding can help move some borderline varieties into the acceptable range for stem length, and also increase yield, since crowding means more plants in a given area.

You can plant varieties that you'd like a longer stem on as densely as possible to maximize yield, keeping in mind that 1) there can be such a thing as too dense; you can't plant everything touching its neighbor like the bulbs we just discussed, and 2) some crops will get diseased if planted too densely.

In particular, foliar diseases can be worse in plants with less air circulation around the leaves. Tighten up spacing slowly on species that are prone to foliar diseases. So, as you get adventurous with cutting down on flower spacing, let stem length and disease be your guide. If the stems are too long or the plants get diseased from lack of airflow, you know they're too close together.

We cobbled together a quarter-acre of tarps from used greenhouse plastic and used landscape fabric to do a combination of solarization and occultation to open ground up for our hemp.
CREDIT: ANDREW MEFFERD

CASE STUDY: GROWING HEMP

U P UNTIL 2019, I had mostly used occultation and solarization to replace rototilling in my greenhouses and fields. This was a fairly smooth transition, especially since we had already been growing in those areas for years; though not perfectly weed-free, there wasn't terrible weed pressure.

However, on short notice in 2019, we got a contract to grow a quarter-acre of hemp. Having just interviewed 20 growers about their no-till systems for my book *The Organic No-Till Farming Revolution*, I wanted to figure out how to

This field had not completely returned to sod in the couple of years since it had last been used. A little over a month of tarping worked pretty well to reopen it.
CREDIT: ANDREW MEFFERD

grow our hemp no-till. I ended up doing for myself what I hope the book will show others how to do: I adapted and changed the methods I had learned about to suit my own particular circumstances. Nobody interviewed for the book was growing hemp, and none of them grew exactly the way that I ended up growing. So, here's what I did.

As our overall planting plan took shape in April, I realized that we would need to plant about a quarter-acre of hemp to satisfy our contract and leave

This photo was taken shortly after the completion of the hemp harvest. I had spent a total of only about ten hours weeding this quarter-acre during the growing season, and you can see there are not that many weeds left after the crop came off the field. The majority of those weeds sprouted from seeds in the compost. Happily, because they were shaded by the crop all season long, many of those weeds didn't have enough time at the end of the season to set seed of their own, so they didn't contribute to the weed seed bank. CREDIT: ANDREW MEFFERD

enough plants for seed production. I had a well-drained quarter-acre field where I had grown vegetables a few years previous but had been fallow since. That seemed like the logical place to put a quarter-acre of hemp. I just had to figure out how to reclaim an overgrown, weedy, not-quite-gone-back-to-sod field without tilling — in just over a month. We were hoping to plant the first week of June.

Before tarping in early May, we put down 500# of lime with a drop spreader. That gave us about a month between tarping and planting, and I knew from researching the book that a month was the bare minimum amount of time needed for occultation to smother the weeds. Though solarization can work in as little as 24 hours on a hot sunny day, I figured our partially regrown sod needed more than that to kill it.

I thought the first step to killing the vegetation would be to do occultation with opaque tarps or solarization with clear tarps. In the end, I did some of both. I looked at our motley assortment of clear and opaque previously

The second year we grew hemp, we solarized the field since we weren't dealing with sod anymore. A month or so of solarization was enough to knock back the light weed pressure from the previous year. Credit: Andrew Mefferd

used plastic and thought: "This will be a good test of whether the clear or opaque works better." When I scraped together every last piece of old landscape fabric and used greenhouse plastic I had, it covered about a quarter-acre. What it ended up being was a test of was how good we are at weighing things down. Almost as soon as we put the tarps down, we got windy storms, and the tarps partially blew off. A big part of the problem was that there were a lot of exposed seams between the tarps. Because I was using randomly sized reused tarps that I had on hand, I had a lot of individual pieces ranging from the width of a greenhouse cover (40 feet) to pieces of landscape fabric that were only a few feet wide.

A complicating factor was that much of the landscape fabric had holes in it where a crop had been planted through it; I soon found that the wind caught this hole-ridden fabric a lot more than an intact piece of plastic. As satisfying as it was to only use materials that I already had on hand, my advice to someone who doesn't have enough material would be to source

In this picture, which was taken just after the tarps were removed, you can see where the edges of the tarps were blown back (the strips of green in the middle). The prevailing winds blow from the direction where the picture was taken, and, because we had reused odd-sized pieces of tarp, the wind got under the edges and blew the tarps partially off. We gradually added more weights as we learned that one of the most important things about tarping is putting enough weights on.
CREDIT: ANDREW MEFFERD

a large used piece of silage cover or greenhouse plastic, with as few seams as possible. Seams are where wind finds a way to get under plastic, so the fewer seams there are, the less likely the cover is to blow away.

My solution to the tarps blowing away was to haul more and more stuff out to the plastic to hold it down. Every time it would blow off, I would haul out more pallets, tires, oddly sized slab wood that wouldn't fit in my stove, cinder blocks, T-posts...anything I had on hand that would weigh a tarp down. Finally, I got enough weight to keep the plastic on. I just wondered how much the weed-killing job would be compromised by all that temporary exposure.

In the intervening month, it rained a lot and was very cool. It was terrible solarizing weather — and not the best for occultation, either. I didn't even get the tarps off until the middle of June, though we had been planning on planting the first week of the month. When I got the tarps off, I was happy with how little vegetation had survived underneath them. There was some regrowth at the edges, where the tarps had temporarily blown off, but mostly the vegetation was dead and starting to decompose.

I did a soil test and, with a drop spreader, put down the required fertilizer which I had amended along the lines of what was recommended for hemp in

my interview with Zach Menchini, "Growing Hemp for the First Time," *Growing for Market*, May 2019 (growingformarket.com).

Many of the no-till growers I interviewed for the book used broadforks to loosen the soil. But with planting already a couple of weeks late, broadforking a quarter-acre was not an option.

I had purchased 80 yards of compost in hopes of layering it 4–6 inches deep and 30 or so inches across the bed top. To apply that much compost, I disengaged the beaters on my old John Deere ground-driven manure spreader, so the moving floor would drop the compost without the beaters flinging it all over the place.

This worked pretty well, but it was slow because the spreader would drop a pile of compost and then drop nothing for a few feet until the floor moved again. This meant that I had to go over each of the 20 or so beds multiple times until they were covered. Better options for spreading would be a high-volume drop spreader or a power take-off [PTO]-driven manure spreader with the beaters disengaged; you could speed up the flow of compost with the PTO

After putting down amendments and 4 inches of compost in the row where the plants would go, we rolled out round bales of spoiled hay in the pathways. Since the beds were planted on 5-foot centers, and the hay bales were 4 feet wide, it left only about a 12-inch strip unmulched where the plants would go.

CREDIT: ANDREW MEFFERD

without increasing ground speed. Partially because of how slow it was, I didn't get around to spreading all 80 yards of compost, so instead of being a layer of compost 4–6 inches deep, it was more like 4 inches deep.

Once the compost was down, we quickly evened it out with rakes, then rolled out some old spoiled round bales of hay in the paths between the plants. Since the round bales were 4 feet wide, and the rows were planted on 5-foot centers, that only left about a foot down the center of each row for weeds to come up. We were hoping that the 4 inches of compost would suppress in-row weeds.

Since I didn't make the compost, I had no idea how weedy it was. I was rolling the dice, hoping it wasn't too weedy. It wasn't too bad; not many weeds sprouted from the compost, except for a few tenacious plants, like milkweed.

Though some parts of the compost were weedier than others. Where our smallest, roughly 4-inch-tall, seedlings were planted into the weedy stuff, the weeds were as big as the hemp plants a month later. I spent about eight hours hand-weeding the worst of the weeds out of the smallest of the plants, and another two hours with a string trimmer whacking the rest of the weeds out from under the bigger plants. I expected that would be the last of the weeding I'd have to do because, once the plants closed canopy, any weeds that grew

from then on would not get any sun.

It is worth noting that field prep like this is still a lot of work! Between our old way of plowing, discing, rototilling, etc. vs. our new no-till way of tarping, re-tarping, removing the tarps, compost spreading, and mulching, it's hard to say which method took less time: tillage vs. no-till. But it doesn't just come down to time spent. I guess a lot of it comes down to what you prefer to spend your time doing. I prefer the jobs associated with tarping to those that go with plowing. You'll need to figure out which jobs you like doing most and what works for your farm — then do that!

This is what our field looked like when it was freshly planted. It's hard to see the plants because they're so small, but they're there.
CREDIT: ANDREW MEFFERD

Right: *This is what our hemp seedlings looked like just after planting.*
CREDIT: ANDREW MEFFERD

Below: *This picture was taken before the hemp closed canopy. You can see how the mulched pathways are clean, and most of the weeds have come up from seeds in the compost. After spending a few hours pulling and weed-whacking this initial flush of weeds, the plants got so big that any weeds that grew from mid-season on were shaded out by the crop.*
CREDIT: ANDREW MEFFERD

By late August, the plants looked like this.
CREDIT: ANDREW MEFFERD

This photo of me was taken shortly before the hemp was harvested. The crop has a closed canopy, so the weeds that managed to survive can't thrive; plus, it was too close to the end of the season for them to go to seed. CREDIT: ANN MEFFERD

Appendix

Tool and Supply Definitions

Mulch: Any material used to cover the soil for any or all of the purposes of: keeping weeds from sprouting, protecting bare soil from hard rain or hot sun, holding moisture, warming the soil or cooling the soil, or enriching the soil (in the case of organic mulches that add organic matter as they break down).

Cover crop: A crop that is not usually sold but planted to improve the soil and provide benefits for a following cash crop. Common benefits include: out-competing weeds, fixing nitrogen (in the case of legumes), keeping living roots in the soil to feed soil life, adding organic matter to the soil, and for use as a mulch grown in place (when the crop is terminated by a roller/crimper for planting a cash crop into).

Cash crop: A crop that is grown to sell. In the scope of this book, we're talking vegetables, flowers, and herbs.

Soil aggregates: Aggregation is what gives soil structure that allows it to resist erosion and let in water and air. Aggregation occurs when soil particles stick together, usually from substances formed by creatures that live in the soil. Aggregation is an important indicator of soil health; it is destroyed by tillage because the aggregates (and the creatures living in them) are crushed.

Sluggo: Since slugs can sometimes proliferate in environments with a lot of residue on the soil, like some no-till systems, it's worth mentioning that Sluggo is an organically approved slug bait and killer. It uses the fact that iron phosphate, which is naturally occurring in the soil, is toxic to slugs. Sluggo comes in little pellets that are irresistible to slugs; when they eat it, they die.

Rotary harrow: A rotary harrow is similar to a rototiller, except instead of the tines rotating around parallel to the soil, the tines rotate perpendicular to the soil, like giant eggbeaters. Though deep rotary harrowing would be as destructive as any other deep tillage, it is possible to set a rotary harrow to a very shallow setting and loosen up a seedbed while at the same time preserving most of the soil structure.

Roller/crimper: Though usually a riding tractor implement, there are small roller/crimpers that can be pulled by walking tractors. They consist of a barrel-shaped roller, which flattens a cover crop. Fins are welded on the outside of the barrel, to press into the ground and crimp the stems of the cover crop as it is being flattened. The crimp prevents moisture from traveling up and down the stem, which is important for terminating the crop. If they are only rolled down without being crimped, vigorously growing cover crops may continue growing.

Broadfork: A wide fork with long tines, designed to be stepped on and rocked gently back and forth as a minimally invasive way of loosening the soil without tilling.

Penetrometer: A tool that senses resistance, which indicates how compacted the soil is. It has a long probe that can be inserted into the soil and a gauge that tells you how much resistance there is in pounds per square inch [PSI]. Most plant roots can't push through soil with resistance over 300 PSI, so if your soil registers higher than that, consider broadforking, adding compost to plant into, or planting soil-loosening cover crops to help with compaction.

Endnotes

1. "No-till and Strip-till Are Widely Adopted but Often Used in Rotation with Other Tillage Practices," https://www.ers.usda.gov
2. Benbrook, C.M. "Trends in Glyphosate Herbicide Use in the United States and Globally," *Environ Sci Eur* 28, 3 (2016), https://enveurope.springeropen.com
3. Daniel Elias, Lixin Wang, and Pierre-Andre Jacinthe. "A Meta-analysis of Pesticide Loss in Runoff under Conventional Tillage and No-till Management," *Environmental Monitoring and Assessment* January 2018, 190:79, https://link.springer.com
4. "Soil Heath Nuggets," https://www.nrcs.usda.gov
5. "The Secret Life of Soil," https://extension.oregonstate.edu
6. Brown, Gabe. *Dirt to Soil*. Chelsea Green, 2018, p. 50.
7. "Black Plastic Tarps Advance Organic Reduced Tillage I: Impact on Soils, Weed Seed Survival, and Crop Residue." *Hortscience* vol. 55(6) June 2020.

Bibliography

Dirt to Soil by Gabe Brown. If you're looking to do no-till with grazing animals, start here. Though the author farms on a large scale in South Dakota, a lot of the principles, like the use of multi-species cover crops and high stocking density of grazing animals, can be scaled to farms of various sizes. His methods can be used as a standalone for a livestock operation, or fields could be rotated between livestock and veg or flower production.

Grow Your Soil! by Diane Miessler. If you want to know more about the soil food web that creates and gives soil life, this book has a very accessible explanation of the relationship between soil life, soil organic matter, and what the critters that populate the soil food web can do for your farm.

The Living Soil Handbook by Jesse Frost. This book dives deep into ways to make your soil healthier and discusses how taking care of your soil will take care of you — with healthier plants.

No-Till Intensive Vegetable Culture by Bryan O'Hara. A master grower describes the systems that keep his farm productive, with a special focus on solarization and high-carbon compost recipes. Though not limited to no-till, this book is an excellent description of overlapping natural systems that can reinforce each other to create a healthy farm.

The No-Till Organic Vegetable Farm by Daniel Mays. This book is an excellent look at the systems on the author's farm, with a special emphasis on how to manage cover crops in a no-till system and how those cover crops can fit into cash crop rotations to make your soil even better.

Organic No-Till Farming by Jeff Moyer. This book is exclusively on the mulch grown in place, roller/crimper method. If you're really interested in this method, this would be a good one to read.

The Urban Farmer by Curtis Stone. Though not strictly a no-till book, the author talks how his farm has transitioned toward no-till methods. The author discusses strategies for running a profitable farm on multiple parcels of borrowed and leased land on a small footprint. An idea whose time has come: we might as well grow the food and flowers close to the people who will buy them.

The Lean Farm by Ben Hartman. Though not specifically about no-till, this book has a lot of great suggestions for cutting out unnecessary steps and making your farm more efficient.

Index

About the Author

A NDREW MEFFERD is the editor of *Growing for Market* magazine and author of *The Organic No-Till Farming Revolution* and *The Greenhouse and Hoophouse Grower's Handbook*. He spent seven years in the research department at Johnny's Selected Seeds, traveling internationally consulting with researchers and farmers on the best practices in organic farming. Before that he worked on the research farm at Virginia Tech, doing field work researching how organic no-till vegetable production compared to tilled organic production. He has worked on farms in Pennsylvania, California, Washington State, Virginia, Maine, and New York State. He now farms in Cornville, Maine.

CREDIT: BY MARY HALEY, MXH MARKETING

Additional Resources from New Society Publishers

The Organic No-Till Farming Revolution
High-Production Methods for Small-Scale Farmers

BY ANDREW MEFFERD

ISBN: 9780865718845
Format: Paperback - 336 pages
Size: 7.5 x 9" (w x h)
Price: US/Can $29.99

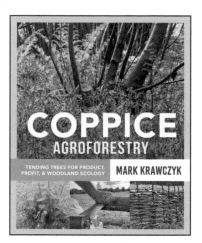

Coppice Agroforestry
Tending Trees for Product, Profit, and Woodland Ecology

BY MARK KRAWCZYK

ISBN: 9780865719705
Format: Paperback - 576 pages
Size: 7.5 x 8.875" (w x h)
Price: US/Can $59.99

Resilient Agriculture, Second Edition

Cultivating Food Systems for a Changing Climate

BY LAURA LENGNICK

ISBN: 9780865719507
Format: Paperback - 368 pages
Size: 7.5 x 9" (w x h)
Price: US/Can $34.99

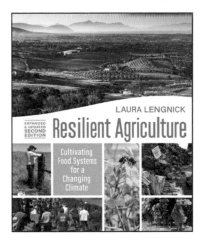

Building Your Permaculture Property

A Five-Step Process to Design and Develop Land

BY ROB AVIS, TAKOTA COEN AND MICHELLE AVIS,

Foreword by GEOFF LAWTON

ISBN: 9780865719378
Format: Paperback - 248 full color pages
Size: 8 x 10" (w x h)
Price: US/Can $49.99

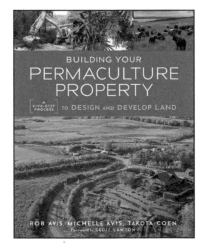

ABOUT NEW SOCIETY PUBLISHERS

New Society Publishers is an activist, solutions-oriented publisher focused on publishing books to build a more just and sustainable future. Our books offer tips, tools, and insights from leading experts in a wide range of areas.

We're proud to hold to the highest environmental and social standards of any publisher in North America. When you buy New Society books, you are part of the solution!

At New Society Publishers, we care deeply about *what* we publish — but also about *how* we do business.

- All our books are printed on **100% post-consumer recycled paper,** processed chlorine-free, with low-VOC vegetable-based inks (since 2002). We print all our books in North America (never overseas)
- Our corporate structure is an innovative employee shareholder agreement, so we're one-third employee-owned (since 2015)
- We've created a Statement of Ethics (2021). The intent of this Statement is to act as a framework to guide our actions and facilitate feedback for continuous improvement of our work
- We're carbon-neutral (since 2006)
- We're certified as a B Corporation (since 2016)
- We're Signatories to the UN's Sustainable Development Goals (SDG) Publishers Compact (2020–2030, the Decade of Action)

To download our full catalog, sign up for our quarterly newsletter, and to learn more about New Society Publishers, please visit newsociety.com

ENVIRONMENTAL BENEFITS STATEMENT

New Society Publishers saved the following resources by printing the pages of this book on chlorine free paper made with 100% post-consumer waste.

TREES	WATER	ENERGY	SOLID WASTE	GREENHOUSE GASES
70	5,600	30	230	30,300
FULLY GROWN	GALLONS	MILLION BTUs	POUNDS	POUNDS

Environmental impact estimates were made using the Environmental Paper Network Paper Calculator 4.0. For more information visit www.papercalculator.org

www.newsociety.com

MIX
Paper from responsible sources
FSC® C016245